D0591114

Encyclopedia of
MAMMALS

Encyclopedia of
MAMMALS

AMY-JANE BEER, PAT MORRIS

Grange
BOOKS

This edition published in 2005 by Grange Books
an imprint of Grange Books Plc
The Grange
Kingsnorth Industrial Estate
Hoo, Near Rochester
Kent ME3 9ND
www.Grangebooks.co.uk

ISBN 1-84013-796-7

Editorial and design:
The Brown Reference Group plc
8 Chapel Place
Rivington Street
London
EC2A 3DQ
UK
www.brownreference.com

Project Director: Graham Bateman
Editors: Angela Davies, Penny Mathias
Art Editor and Designer: Steve McCurdy
Cartographic Editor: Tim Williams
Editorial Assistants: Marian Dreier, Rita Demetriou
Picture Manager: Claire Turner
Picture Researcher: Vickie Walters
Production: Clive Sparling
Researchers: Dr. Erica Bower, Rachael Brooks,
Rachael Murton, Eleanor Thomas

Printed in Singapore
1 2 3 4 5 08 07 06 05 04

North American river otter

Title page: **Grizzly bear**
Half title: **Bull elk**

Contents

House mouse

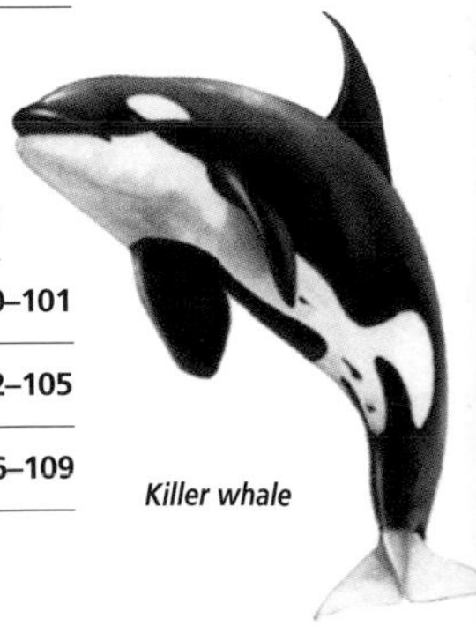

Killer whale

Common raccoon

Walrus

Reindeer

Introduction

OUR PLANET is home to approximately 5,000 species of living mammals that display enormous variety. To see them all in the wild you would have to visit the shores of oceans, travel from Arctic tundra to subtropical swamps, and explore rainforest, open prairie, barren desert, high mountains, tropical grasslands, and city sewers. Many bats or mice could fit easily in the palm of a toddler's hand. A large polar bear on the other hand, stands 10 feet tall on two legs and weighs as much as ten men, while the blue whale is the largest animal ever to have lived on earth.

Some mammals are held up as symbols of nobility and strength; others are reviled as vermin or feared as predators. Loved or loathed, few are regarded with ambivalence—one way or another, mammals always seem to elicit an emotional response. Perhaps this is because as mammals ourselves we can at least partially relate to their view of the world—we experience the same drives to obtain food and shelter and protect ourselves from attack. We, too, compete with others of our kind but sometimes rely on close alliances with others to survive. Above all, we, too, go through an extended period of parental care and experience the same urge to protect and provide for our young, which like all mammals, begin life dependent on milk produced by their mother in mammary glands. Like all mammals our bodies are covered (albeit sometimes sparsely) in hair, and our bodies are warm. It is this ability to generate body heat internally (by metabolizing food), that has allowed mammals to populate nearly every habitat on earth, from desert to pole.

Mammal Evolution

MAMMALS EVOLVED from reptile-like ancestors about 200 million years ago. Mammals developed from a group of mammal-like reptiles called cynodonts. While dinosaurs dominated the land, these warm-blooded, furry creatures remained small and ratlike. It was not until the early Eocene period about 50 million years ago that they began a rise to prominence. The extinction of the dinosaurs around 65 million years ago had allowed mammals to grow larger, and mammals have since come to dominate the earth, just as dinosaurs did before them.

On land the mammals quickly diversified. In the northern hemisphere the first horses, camels, elephants, monkeys, and rodents appeared, along with the forerunners of carnivores such as wolves, bears, and cats. These were all placental mammals. Their young developed inside their mothers and were born well developed. In South America, Australia, and Antarctica a different group, the marsupials, flourished. The ancestors of all marsupials appeared in North America about 120 million years ago. Most have since died out in that continent, leaving the Virginia opossum as the group's sole native representative. A newborn marsupial is born incompletely developed. It continues to develop outside the uterus, attaching itself to the mother's body near her mammary glands. Some marsupials have a pouch that shelters the suckling young. A few mammals, such as spiny anteaters, reveal their reptile origins by laying eggs.

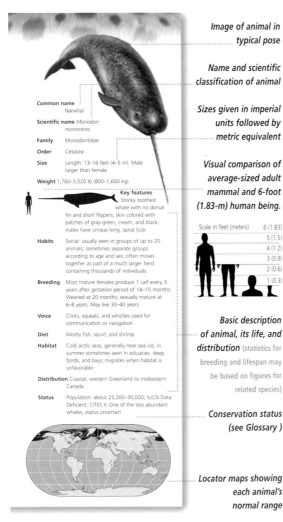

Image of animal in typical pose

Name and scientific classification of animal

Common name
　　　Narwhal

Scientific name *Monodon monoceros*

Family　Monodontidae

Order　Cetacea

Size　Length: 13–16 feet (4–5 m). Male larger than female

Weight 1,760–3,520 lb (800–1,600 kg)

Key features
Stocky toothed whale with no dorsal fin and short flippers; skin colored with patches of gray-green, cream, and black; males have unique long, spiral tusk

Habits　Social: usually seen in groups of up to 20 animals; sometimes separate groups according to age and sex; often moves together as part of a much larger herd containing thousands of individuals

Breeding　Most mature females produce 1 calf every 3 years after gestation period of 14–15 months. Weaned at 20 months; sexually mature at 6–8 years. May live 30–40 years

Voice　Clicks, squeals, and whistles used for communication or navigation

Diet　Mostly fish, squid, and shrimp

Habitat　Cold arctic seas, generally near sea ice; in summer sometimes seen in estuaries, deep fjords, and bays; migrates when habitat is unfavorable

Distribution Coastal; western Greenland to mideastern Canada

Status　Population: about 25,000–30,000; IUCN Data Deficient; CITES II. One of the less abundant whales, status uncertain

Sizes given in imperial units followed by metric equivalent

Visual comparison of average-sized adult mammal and 6-foot (1.83-m) human being.

Scale in feet (meters)　6 (1.83)
　　　　　　　　　　　5 (1.5)
　　　　　　　　　　　4 (1.2)
　　　　　　　　　　　3 (0.9)
　　　　　　　　　　　2 (0.6)
　　　　　　　　　　　1 (0.3)

Basic description of animal, its life, and distribution (statistics for breeding and lifespan may be based on figures for related species)

Conservation status (see Glossary)

Locator maps showing each animal's normal range

← Summary panel presents basic facts and figures for each mammal.

By around 50 million years ago bats flew in the night sky. Bats had a dramatic effect on moths and other nighttime insects. So severe was their predation that one group of moths gave up nocturnal life altogether and switched to the daytime. These insects became modern butterflies.

By the Miocene (23.7 to 5.3 million years ago) hoofed mammals such as deer and pigs thrived, while rodents diversified to become the largest of mammal groups. Being warm-blooded helped mammals survive climatic changes during the Pliocene, when long, cold periods called ice ages were interspersed with warmer spells.

Humans and Other Mammals

HUMANKIND HAS undoubtedly been responsible for the loss of many mammal species. In more recent times, however, we have also been able to save some species that were facing imminent extinction. Among the first mammals to benefit from a new-found sense of responsibility for fellow animals were the American beaver and bison. The gray whale, too, was saved by an early ban on hunting. But bison and whales are very large, highly conspicuous animals. Their precipitous decline over a relatively short period of time could not help but be noticed. The same cannot be said for many smaller species facing a similar fate today. You may have heard of the plight of the black-footed ferret or the black-tailed prairie dog, but what about the West California kangaroo mouse or the Nevada chipmunk? Without a significant conservation effort these and many other little known species face extinction in our lifetime. Our responsibility to these smaller species is no less than to the majestic bison, the extraordinary pronghorn or the great gray whale. It is only by coming to understand these animals, all fascinating examples of natural history, that we have a hope of offering them the secure future they deserve.

The Encyclopedia of Mammals

IN THIS ENCYCLOPEDIA you will find detailed descriptions of 79 common or familiar mammals. Examples have been selected to give the broadest range of types from distinctive habitats. For each animal there is a detailed summary panel (see left) that gives all the basic facts and figures, including a world distribution map and a scale drawing compared with a 6-foot-high person. There then follows the main article, which describes the most interesting features of each animal. Throughout there are detailed artwork portrayals and dynamic photographs of the animals in the wild.

If you find something in the following pages to inspire you, we hope you will pass it on and thus contribute in your own way to a greater awareness and appreciation of our planet's magnificent mammal life.

⊕ *Coat color variations in the northern red fox.*

Bobcat

Felis rufus

Territorial and solitary, the bobcat is sometimes confused with its close cousin, the lynx. Both cats have tufted ears and short tails, but the bobcat tends to be the more aggressive of the two.

Common name Bobcat

Scientific name *Felis rufus*

Family	Felidae
Order	Carnivora
Size	Length head/body: 25.5–41 in (65–105 cm); tail length: 4–7.5 in (11–19 cm); height at shoulder: 17.5–23 in (45–58 cm)

Weight 9–33 lb (4–15 kg)

Key features	Small, slender-limbed, short-tailed cat; fur thick, varies in color from buff to brown with darker spots and streaks; ears pointed, often with tufts; ruff of fur around jowls
Habits	Solitary; territorial; active day or night
Breeding	Litters of 1–6 kittens born after gestation period of 60–70 days, usually in spring. Weaned at 2 months; females sexually mature at 1 year, males at 2 years. May live up to 32 years in captivity, probably no more than 13 in the wild
Voice	Usually silent, but hisses and shrieks in distress and during courtship
Diet	Small mammals and birds; sometimes larger prey, such as small deer; domestic animals
Habitat	Varied; includes forests, scrub, swamp, mountains, and the edges of deserts

Distribution North America

Status	Population: 700,000–1 million; CITES II. Declined in the past due to persecution; still harvested for fur under license in some states

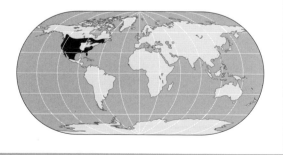

THE BOBCAT IS SO CALLED because of its short tail, which resembles the docked bobtails of some domestic mammals. It looks a lot like another North American cat, the lynx, but there are few places where the two species live alongside each other. Bobcats are more aggressive than lynx and usually drive the latter out of habitats that suit both. However, the lynx is much better adapted to snow than the bobcat, which has small feet that sink in easily. So the northern limit of the bobcat's distribution is determined largely by the average snowfall.

Adaptable Cats

Bobcats are true generalists, which means they can live in almost any habitat; hence their wide natural distribution throughout most of North America. They are only absent where large areas are intensively cultivated or given over to industrial development. They are scarce in places where they have been overhunted. Their varied diet is a major factor in their adaptability. Their preferred prey appears to be rabbits and hares, but they will eat other small mammals and many larger ones too, including beavers, peccaries, and deer. Hoofed mammals are the main winter prey of bobcats in the north of their range, and Canadian bobcats are usually larger than those living in the south. This helps them cope with bigger prey.

The bobcat's hunting technique almost always relies on surprise. With its mottled coat providing admirable camouflage, a bobcat can sneak up on the most alert of victims, using a combination of stealth and endless patience. The kill is made with a sudden leap and a quick bite to the back of the neck, separating the backbones and severing the spinal cord.

Bobcats can be active at any time of day, but most animals adjust their activity to match that of their preferred prey. They wander up to 9 miles (15 km) a day in search of food, stopping often to mark and re-mark the boundaries of their home range. Females have smaller ranges than males—0.4 to 8 square miles (1 to 20 sq. km)—but they do not overlap with any others. Male territories can be anything from 2 to 16 square miles (5 to 40 sq. km), and they can overlap the ranges of other males and several females.

Respecter of Boundaries

Outside the breeding season bobcats go out of their way to avoid meeting, which leads to intensive scent marking to warn others away. Marks are made with urine and feces, and with secretions from the cat's anal glands. The marking is very effective, and bobcats appear to respect each other's territorial boundaries. Aggressive encounters seem very rare, and the ownership of a particular range area only changes when the resident animal dies.

⊕ Bobcats are solitary animals. Outside the breeding season they will go out of their way to avoid meeting and seem to respect each other's territories.

Of course, the need to breed means that males and females must meet at some point, and mating occurs any time between November and August. Most kittens are born in spring, but some births happen much later in the year. If a female loses her first litter of the year when the kittens are very young, she comes into season again and may produce a replacement litter in late summer. The kittens are able to follow their mother after three or four months, and they learn hunting skills by watching her. They stay with her until she is ready to breed again, then head off to find a place of their own.

There are probably about 1 million bobcats living in North America. They are protected in some states, notably those where the species has become rare. Elsewhere, they are hunted and trapped for part of the year, and their pelts sold to the fashion industry.

Common name Lynx (Eurasian lynx)

Scientific name *Felis (Lynx) lynx*

Family	Felidae
Order	Carnivora
Size	Length head/body: 31–51 in (80–130 cm); tail length: 4–10 in (10–25 cm); height at shoulder: 23.5–29.5 in (60–75 cm)
Weight	18–84 lb (8–38 kg)
Key features	Stocky cat with longish legs and large, furry feet; color varies from pale gray through yellow to reddish-brown; ears tufted
Habits	Solitary; nocturnal; wanders widely
Breeding	Litters of 1–4 kittens born April–June after gestation period of 67–74 days. Weaned at 3 months; females sexually mature at 9–21 months, males at 21–31 months. Lives up to 24 years in captivity, 17 in the wild
Voice	Hisses and mews, but usually silent
Diet	Mostly eats small- and medium-sized mammals, including hares and small deer
Habitat	Mixed and taiga forest, scrub, steppe, rocky alpine slopes
Distribution	Eurasian lynx: northeastern Europe, Balkans, Turkey, and the Middle East excluding Arabia, much of former U.S.S.R., Mongolia, and northern China. Iberian lynx: Spain, Portugal. Canadian lynx: Canada, Alaska, northern U.S.
Status	Population: unknown, but certainly many thousands; IUCN Endangered (Iberian), Vulnerable (Canadian); CITES I (Iberian), II (Canadian and Eurasian). All have declined, mainly as a result of hunting for fur

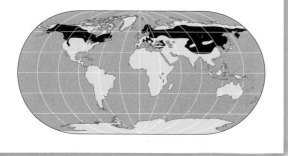

Lynx

Felis lynx

A stocky, medium-sized cat, the lynx is widely distributed throughout the Northern Hemisphere. It is sometimes considered to be three separate species.

THERE IS AN ONGOING SCIENTIFIC debate about whether the three recognized types of lynx are all members of the same species or not. The Canadian lynx (*F. canadensis*), Iberian lynx (*F. pardinus)*, and Eurasian lynx (*F. lynx*) look remarkably similar, but they have different behavioral adaptations to suit life in their different parts of the world.

Distinctive Tail

Lynx are close cousins of bobcats, but can be told apart by examining the tail. Both species have short tails, but that of the lynx is completely black at the tip. In contrast, the bobcat's tail tip is black just on top. The largest lynx are Eurasian specimens from Siberia. They live on Arctic hares and other mammals several times bigger than themselves, such as reindeer (caribou). Snow can be an advantage to a hunting lynx, since deer can become bogged down and are then easier to catch. The lynx's feet are large and furry, so its weight is spread over a larger area, allowing it to run across snow without sinking. Iberian and Canadian lynx are about half the size of Eurasian lynx and generally hunt smaller prey.

Feeding Habits

The Canadian lynx feeds almost exclusively on snowshoe hares, and its numbers fluctuate from year to year according to the availability of the hares. Iberian lynx mainly feed on mammals such as rabbits, although they are also able to catch birds and fish, hooking them out of the air or water with a swipe of their sharp claws. For the smaller lynx a rabbit a day is sufficient food, but larger lynx eat rather more. Having killed a big animal such as a deer, they will drag it to safety, eat what they can, and cache (store)

⊖ Snowy conditions can be advantageous for the hunting lynx, since its large, furry feet help spread its weight evenly and stop it from sinking into fresh snow. Lynx will often hunt deer that get bogged down in the snow and so are relatively easy to catch.

the rest for later. Hunting is almost always a solitary activity, although mothers have sometimes been seen helping their fully grown young to hunt. Newly independent lynx sometimes team up with a sibling for the first few months after leaving their mother's care.

Endangered Species

Female lynx mature faster than males and can be capable of breeding within their first year. However, few do so because breeding is regulated by habitat availability. Lynx do not breed until they have found a suitable home range in which it will be possible to rear young. In places like Spain, where habitat is greatly restricted, adult lynx may never get the opportunity to breed. Of the few hundred Iberian lynx left in the wild fewer than a third are thought to be breeding females, making this one of the world's most endangered cats. Canadian and Eurasian lynx are faring better, although both have been extensively hunted in the past. Lynx fur is dense and luxurious, and several thousand animals are still legally shot or trapped every year for their fur.

In Central Europe lynx have been reintroduced to parts of Germany, Slovenia, and Switzerland; and while it is still early days for these cats, there are encouraging signs. The Swiss animals have bred successfully for several seasons, and some have now spread over the Alps into northern Italy of their own accord.

Common name
Puma (cougar, panther, mountain lion, catamount)

Scientific name *Felis concolor*

Family Felidae

Order Carnivora

Size Length head/body: 38–77 in (96–196 cm); tail length: 21–32 in (53–82 cm); height at shoulder: 24–27.5 in (60–70 cm)

Weight Male 148–264 lb (67–120 kg); female 80–132 lb (36–60 kg)

Key features Large, muscular cat with long legs and tail; small head with large, rounded ears; coat color varies from silvery gray through warm buffy tones to dark tawny

Habits Solitary; active at any time of day; climbs extremely well

Breeding Litters of 1–6 (usually 3 or 4) kittens born January–June after gestation period of 90–96 days. Weaned at 3 months; sexually mature at 2.5–3 years. May live up to 21 years in captivity, rarely more than 14 in the wild

Voice Hisses, growls, whistles, and screams

Diet Carnivorous; mostly deer; also other hoofed animals, rodents, and hares

Habitat Very varied; lowland and mountain forests, swamps, grassland, and scrub

Distribution Most of North and South America

Status Population: many thousands in total, but Florida panther (*F. c. coryi*) fewer than 50; IUCN Critically Endangered (2 subspecies); CITES II (at least 2 subspecies). Persecuted as a pest in the past; now protected in parts of its range although still hunted in other areas

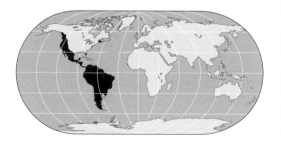

Puma

Felis concolor

The puma is the second largest cat in the Americas and by far the most widespread, with a natural range extending from Canada to Patagonia.

PUMA, COUGAR, PANTHER, AND MOUNTAIN lion are widely used names for the same animal—a highly adaptable, agile predator that feeds on medium-sized prey such as deer. Despite their larger size, pumas are more closely related to lynx and bobcats than to lions and jaguars, and are first cousins to the domestic cat. They are extremely agile and can climb with great ease. They prey mostly on ground-dwelling animals, but often use trees to lie in wait for passing animals, dropping on them from above. Alternatively, they may chase a prey animal for a short distance before leaping on its back. In either case the prey is killed with a bite to the neck. A lone adult puma may only need to kill every two weeks. It will drag the carcass to a safe place and hide it under a heap of dirt and debris, returning to feed on it again and again. For a mother puma with cubs life is rather more demanding, and she may have to kill a deer every three or four days to sustain her family.

Solitary Existence

Pumas are generally solitary, although young cats may stay with their mother for over a year and then remain together a few more months after she has left them. After the family disperses, young pumas live as nomads for a while, wandering through the ranges of resident pumas until they find a place to settle. While they may be capable of breeding by the age of two years, they will not do so until they have established themselves in a suitable home.

Females occupy large home ranges, which may overlap more or less completely with those of other pumas, but they avoid meeting by the use of scent marks and various vocalizations. Except when they have young kittens, females wander widely over their entire range, using

⊕ Pumas have a reputation for killing livestock such as sheep and cattle. In fact, they more frequently kill deer and tend to select old or weak individuals. In so doing they may be helping maintain a healthy deer population.

various patches of dense vegetation or small caves to rest in, rather than a regular den. Males operate in a similar way, except their ranges are much larger—sometimes over 400 square miles (1,000 sq. km)—and they overlap only with female pumas, not other males. They use scent marking more frequently than females, especially around the borders of their range. They do not generally fight over territory, and new residents only move in when the previous occupant dies.

Pumas have a reputation for killing livestock such as horses, cattle, and sheep, but that is relatively infrequent. They kill deer, too, but in so doing may actually help keep the deer population healthy, since they tend to select old or weak individuals. It also prevents the deer from getting too numerous. Pumas have been implicated in a number of fatal attacks on humans, but in general they avoid people.

Gradual Comeback

Intensive eradication attempts all but exterminated pumas from much of North America, leaving only small populations in the western mountains, southern Texas, and Florida. The animals appear to be making a gradual comeback in some Midwestern and eastern states, but they are still hunted in Texas. The Florida population is thought to number no more than 50 individuals, despite millions of dollars being spent on their conservation.

Common name Gray wolf (timber wolf)

Scientific name *Canis lupus*

Family	Canidae
Order	Carnivora
Size	Length head/body: 35–56 in (89–142 cm); tail length: 12–20 in (30–51 cm); height at shoulder: 23–28 in (58–77 cm)

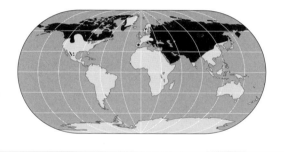

Weight 22–175 lb (10–80 kg). Male larger than female

Key features Large, long-legged dog with thick fur and bushy tail; fur usually gray, although color varies with distribution

Habits Social, although sometimes solitary; more or less nocturnal; hunts communally to bring down prey up to 10 times its own weight

Breeding One to 11 (average 6) pups born in a den after gestation period of 63 days. Weaned at 5 weeks; sexually mature at 2 years. May live up to 16 years in captivity, rarely more than 13 in the wild

Voice Growls, barks, whines, and howls

Diet Mainly large mammal prey, including deer, moose, muskox, mountain sheep, bison, beavers, and hares

Habitat Almost anywhere from tundra to scrub, grassland, mountains, and forest

Distribution Northern Hemisphere

Status Population: many thousands; IUCN Vulnerable; CITES I (India, Pakistan, Nepal, Bhutan); elsewhere CITES II. Now more stable following centuries of persecution

Gray Wolf

Canis lupus

Wolves are intelligent and adaptable creatures, often living in close-knit family groups. Human attitudes to wolves range from deep respect to outright hostility, fueled by chilling folk tales of their wickedness.

THE GRAY WOLF IS THE LARGEST species of dog. It once lived throughout the Northern Hemisphere in all but the most extreme tropical and desert habitats. Only one other mammal has a greater natural range or lives in a wider variety of habitats—our own species. Persecution by humans led to a dramatic decline in wolf numbers worldwide over the last 300 years, and the species has become extinct over much of its former range. It is now associated only with areas of wilderness. Wolves disappeared altogether from Britain in the 18th century and from Japan and much of western Europe in the following 200 years.

Eradication Program

In North America the gray wolf was the chief target of a prolonged campaign of predator eradication that began soon after the arrival of European settlers. Wolves were shot and trapped in such numbers that by 1940 there were none left in the western United States, and numbers elsewhere were in serious decline. More recent methods of control include poisoning and sport hunting from aircraft. Similar eradication programs in the former Soviet Union reduced wolf numbers there by about 70 percent. In other parts of Asia the wolf is now rare. The Mexican wolf is officially listed by the IUCN as Extinct in the Wild, with only about 140 remaining in captivity.

More recently, however, studies of wolf populations have convinced biologists that far from being a scourge of the land, wolves are in fact an important stabilizing influence on wilderness ecosystems. Such discoveries, along with a growing sense of responsibility toward wildlife in general, have prompted several wolf

⊙ Gray wolves from different geographical areas may vary in size and appearance. Those living in arctic and mountainous regions, for example, are much larger than their relatives in the hot, dry scrublands of Arabia.

Folklore: Who's Afraid of the Big, Bad Wolf?

Wolves have long been the subject of myths and legends. Stories such as *Little Red Riding Hood* and *The Three Little Pigs* cast wolves as cold-blooded killers of men and domestic animals. On the other hand, the legend of Romulus and Remus —the babies raised by wolves— and Kipling's *Jungle Book* stories portray wolves as wise and devoted parents. In reality the wolf is all these things and more.

conservation projects around the world. Several European populations have now been saved from extinction, and the range of the wolf in North America is increasing slowly. In most places where man and wolf still live side by side there is now an uneasy truce enforced by laws protecting the wolf from direct persecution, but giving livestock owners some rights to protect their property. Nevertheless, many country people are not happy to share their land with wolves and want them shot or trapped. Efforts to reintroduce wolves to Yellowstone National Park have also run into difficulties with hostile residents in surrounding areas.

Geographical Differences

Not surprisingly for such a widespread species, wolves from different geographical areas vary considerably in size, appearance, and behavior. The biggest wolves live in large packs in the tundra regions of Canada, Alaska, and Russia. Their relatives in the hot, dry scrublands of Arabia are smaller and more likely to live alone or in small groups.

The size of a wolf pack is controlled largely by the size of its most regular prey. Lone wolves do well where most of their food comes from small prey, carrion, or raiding human refuse. Where deer are the main prey, packs of five to seven animals are usual.

However, pack sizes may be larger still where wolves feed on very big prey. In the Isle Royale National Park in Lake Superior, for example, where the animals feed almost exclusively on moose, packs may include more than 20 animals.

Selective Predation

Wolves normally hunt old, young, weak, or disabled prey and soon give up an attack if the animal is able to defend itself or make a quick getaway. In fact, only about 8 percent of wolf hunts end in a kill, which is why it is highly unlikely that wolf predation does any real harm to prey populations as was once feared.

A large wolf needs to eat an average of 5.5 pounds (2.5 kg) of meat every day, but will often go for several days without food. However, when a kill is made, it makes up for any such lean periods by "wolfing" up to 20 pounds (9 kg) in a single meal. A large prey animal may keep a pack well fed for several days. During the time they are not actively feeding, the wolves may rest near the carcass to defend it from scavengers.

Wolf attacks on humans are rare. In North America, for example, there are no fully documented cases of unprovoked attacks on people by healthy wolves. However, wolves can and do attack livestock. Sheep and cattle are, after all, close relatives of the wolf's natural prey and yet far easier to catch and kill because generations of domestication have made them virtually incapable of defending themselves. They are large, meaty, and prone to panic, and are often penned in with no hope of escape. Even so, it would rarely take more than the sight of a human to cause the wolves to abandon the hunt and run away.

Sibling Care

A wolf pack is made up of a single breeding pair and their offspring of the previous one or two years. The nonbreeding members of the pack are usually young animals. They are prevented from breeding by the dominant pair, but help care for their young siblings. In areas where good wolf habitat is plentiful, young

⬆ *The size of a wolf pack is usually determined by the size of available prey. For example, when deer are the main food source, packs of five to seven are common. Wolves preying on larger animals, such as moose, often belong to packs of 20 or more.*

Reintroduction

In 1995, after years of careful planning and much controversy, 31 Canadian-born gray wolves were released into Yellowstone National Park. The park contains over 17 million acres (7 million ha) of prime wolf habitat and also supports large herds of elk. The relocated wolves have thrived since their introduction, as have those released in other locations in Montana and Idaho. The interests of local ranchers are protected in that they are compensated for wolf attacks on their livestock. In addition, farmers are now permitted to shoot wolves on their own land. In the first four years of the program nine wolves were shot legally.

The success of the Yellowstone project has encouraged conservationists to consider reintroducing the wolf elsewhere. One highly controversial plan is to release captive-bred wolves in Scotland, a country that has not seen wild wolves for 300 years. The problem with the idea is that islands where wolves could be out of the way of humans are too small to support a viable population. Yet on the mainland there are too many people and sheep for the wolves to live without causing trouble.

A gray wolf pup at the entrance to its den. On average, a litter contains about six pups.

wolves may leave their parents' pack as early as 12 months of age. Some stay with the family for a further season; but by the time they are fully mature at 22 months, they will move on. Dispersing animals may live on the edge of their parents' territory until a suitable mate comes along. Other young wolves scatter widely in search of a mate and territory of their own.

Territorial Howling

The pack occupies a territory of anything from 8 to 5,200 square miles (20 to 13,000 sq. km), the exact size varying according to the number of wolves and quality of habitat. All pack members help defend the territory, and they will travel to every part of it at least once a month, moving in single file along regular routes. They mark their territory with scents, scratches, and long sessions of howling. In open country wolf howls can be heard up to 10 miles (16 km) away, even by human ears. When wolves from neighboring packs do meet, the

encounters often lead to serious fights in which one or more animals may be fatally wounded. To minimize the risk of such incidents, the wolves usually leave a kind of buffer zone of seldom-visited land around the edge of their territory. Such areas also serve as a kind of reservoir for prey, which is only exploited in times of food shortage.

All wolves are highly adaptable. While the social structure of a pack may stay the same for many years, individuals are able to switch roles with surprising ease. The dominant (or alpha) male leads the pack and is responsible for initiating hunts or other movements. If he dies or is absent for long periods, the alpha female takes on the leadership role until a new alpha male moves in. Subordinate wolves rise to dominance almost as soon as an alpha wolf dies, and both sexes are capable of rearing older cubs on their own if their mate dies.

↑ *Wolves communicate using body language and facial expressions. Above, a defensive threatening posture (1); a submissive greeting (2); and an offensive threatening pose (3).*

17

Coyote

Canis latrans

Opportunistic and resilient, the North American coyote is the archetypal predator. The species continues to thrive throughout its range, despite centuries of persecution by humans.

Common name Coyote

Scientific name *Canis latrans*

Family	Canidae
Order	Carnivora
Size	Length head/body: 30–39 in (76–100 cm); tail length: 12–19 in (30–48 cm); height at shoulder: about 24 in (60 cm)
	Weight 15.5–44 lb (7–20 kg). Male slightly larger than female
Key features	Typical wolf but smaller and slighter in build than gray wolf; ears large and pointed; muzzle narrow; fur shaggy and usually a shade of beige or gray; paler on belly, but darkening to black on tip of tail
Habits	Mostly nocturnal, but can be active at any time of day; some migrate into mountains in summer; less social than gray wolf
Breeding	Litters of 2–12 (average 6) born in spring after gestation period of 63 days. Weaned at 5–6 weeks; sexually mature at 1 or 2 years. May live up to 21 years in captivity, usually fewer than 15 in the wild
Voice	Wide repertoire of barks, whines, and howls
Diet	Carnivorous; mostly mammals, including rabbits, woodchucks, rodents, and deer; also carrion
Habitat	Grasslands and prairie, scrub, and forest
Distribution	North America
Status	Population: abundant. Common and widespread; hunted for fur and as a pest

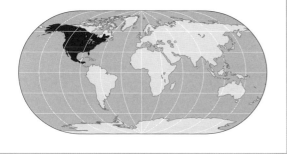

THE COYOTE IS ONE OF THE WORLD'S most successful carnivores. The species occupies a vast range, and local populations seem able to adapt to a wide variety of habitats. The size of the animals, their diet, and their social structures are all flexible in order to make the most of different environmental conditions wherever they live. Coyotes continue to do well despite centuries of intensive persecution by humans.

Coyote Persecution

Millions of coyotes have been killed for their fur and to protect game and livestock, especially sheep. Young coyotes are killed in their dens, and the adults are trapped or shot by marksmen on foot or in aircraft. Poisoning used to be a major method of coyote control. It was outlawed in 1972, partly because it was considered cruel, but also because many other species were harmed accidentally by eating poison meant for coyotes. Coyote predation still costs farmers millions of dollars a year. While there is no evidence that the population is seriously threatened by the ongoing persecution, the control of coyotes has become highly controversial. Recent investigations suggest that in most states coyote predation on livestock is not as common as people thought.

Ironically, the arrival of European settlers in North America has done more to extend the coyote's range than to control its numbers. Before human settlement coyotes were restricted to the plains of central North America by the lack of suitable habitat elsewhere and the presence of wolves, which were bigger and better adapted to forest life. As the human population expanded westward, landscapes changed. The forests were felled and replaced

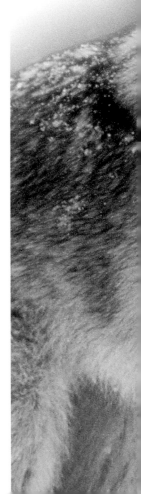

⊕ *A coyote howls in a snowstorm in Yellowstone National Park. Its long howl is only one of a varied repertoire of calls.*

with pasture and arable land. Human fear of wolves meant that the coyote's main competitor was eradicated from many states in a short space of time. The coyote was presented with an unprecedented opportunity for expansion, and today the species occupies all but the extreme northeast of the continent.

Hybridization

In places where advancing coyotes encountered remaining populations of gray and red wolves, the species interbred. The resulting hybridization almost proved disastrous for the red wolf, purebred examples of which became so rare that in 1975 the remaining population had to be taken into captivity for its own protection. Unfortunately, after 25 years of captive breeding and subsequent release onto coyote-free reserves, there are still fewer than 300 red wolves left in the world.

By contrast, coyotes are so numerous that interbreeding has done little to dilute their gene pool. In fact, evidence suggests that an injection of wolf genes has resulted in larger coyotes in Canada better adapted to life on the tundra. Even where there are no wolves, coyotes have gotten bigger, partly as a

The Coyotes Come to Town

Coyotes are adaptable animals. They are capable of living in a wide variety of habitats, including suburban areas. As cities expand, many animals die out, being unable to adjust to the new conditions. Not so the coyote! The species is often seen in the outskirts of cities like Denver, Houston, and Boise, and is particularly familiar in Los Angeles. It appears that coyotes did not invade Los Angeles, but merely stayed put as the city spread around them. Los Angeles is especially suitable because there are many scrub-filled ravines, large gardens, and other relatively undisturbed areas where they can live, emerging to feed on trash, food scraps, and other urban animals. The red fox has made a similar success of urban living in parts of Britain.

result of improved food resources. Coyotes living in the Mexican desert average about 28 pounds (13 kg). Those that have colonized Alaska regularly exceed 42 pounds (20 kg).

The coyote is smaller than a wolf, but substantially bigger than a fox. It can be difficult to distinguish from a wolf, but the narrow snout, long ears, and small feet are identifying features. Another useful clue, especially at a distance, is the tail. It is long and brushy like that of most other wild dogs. However, when running, the coyote carries its tail in a low sweep, not high like a wolf or straight out like a fox, nor curled like some domestic dogs. Other less obvious differences are clues to the coyote's lifestyle. A coyote's skull has a pronounced central ridge running from front to back (called a sagittal crest), which allows for the attachment of powerful jaw muscles that are much bigger than those of a fox. The coyote has crushing molar teeth and long, pointed canines, designed for tearing and chewing chunks off large prey. Foxes, on the other hand, have more pointed teeth, less muscular jaws, and feed on smaller animals.

Coyotes are out-and-out carnivores, with the flesh of mammals providing at least 90 percent of their food. The exact composition of their diet varies with habitat and season, from rabbits and rodents on grassland to mostly deer in the forests of Minnesota. Some coyotes have learned rudimentary fishing techniques, while others occasionally catch and eat birds. Fruit and vegetables are eaten in season. Hunting techniques vary depending on the prey, but a large proportion of the animals caught are old, sick, or immature. Small prey are stalked and pounced on from above, while large animals may be chased over long distances. The coyote is one of the fastest predators in the Americas, able to run down prey at speeds of up to 40 miles per hour (64 km/h).

Hunting as a pack is a definite advantage when chasing large prey over a distance, since different members take turns leading the chase until the quarry tires. Coyotes tend to live in packs only where large prey animals, such as deer, are concentrated in an area. Packs usually consist of three to seven closely related animals. Larger groups sometimes gather around a temporary food source, such as a large carcass, but are not assembled for long. In less productive habitats coyotes live in pairs or alone and hunt small animals over a wide area.

⊕ *A coyote pack defends a carcass on the edge of its territory. Three pack members (1) feed while the dominant male (2) threatens an intruder (3), who assumes a defensive threat posture. Another male (4) backs up his leader, but shows less aggression. Another trespasser (5) looks on while other coyotes (6) wait in their own territory for the pack to leave.*

Hunting Partnerships

Cooperative hunting is not unusual among carnivores—lions do it, and so do some species of otters and dogs. Coyotes sometimes hunt in packs, but by far their most remarkable teamwork is performed with an unlikely partner: the American badger. The two animals form an alliance and use their combined skills to catch prey that would otherwise escape them. The coyote sniffs out small burrowing mammals under the soil and waits patiently while the badger uses its powerful forelimbs and huge claws to dig them out. The badger's sense of smell is not good enough to detect buried prey, and the coyote's small feet mean that digging is a slow and laborious exercise. Once caught, the prey is shared amicably between the partners. Cooperation between the two animals has evolved over generations and works because both benefit equally.

Male coyotes have large home ranges of up to 32 square miles (80 sq. km), which can often overlap with those of other males. Female territories rarely exceed 6.5 square miles (17 sq. km), but are generally exclusive, each female having her own territory. Coyotes seem to choose obvious landmarks such as streams and tree lines to define their territories, and they mark them with scent in urine and feces.

Courtship

A single female may be courted by several males over a period of two or three months in spring. However, once she has chosen one male to be her mate, the relationship may last several years, sometimes for life. The female bears just one litter a year, but it can include 10 or more pups (litters of 19 pups have been recorded, but it is highly unlikely that all could survive).

The average litter contains five to seven pups. In good habitat pups are tended by both parents and one or more elder sisters. Helpers only tend to stay with their parents where there is no shortage of large prey, and so their presence may be more important in defending the young and the den than in obtaining food.

Coyote pups begin to eat regurgitated meat at just three weeks old, and by the time they are six weeks old they no longer need their mother's milk. They put on weight fast and are fully grown by the age of nine months. Male offspring disperse at this time, while females may stay behind for a further two or three years. Dispersing coyotes travel an average of 18 miles (30 km) from the den where they were born, but tagging studies show that some travel hundreds of miles before settling down to raise a family of their own.

⊕ As a form of greeting, coyotes will often rear up on their hind legs and nuzzle each other's face. Aggressive encounters begin in a similar way, but may develop into a wrestling match, with rolling and biting.

Red Fox

Vulpes vulpes

The red fox is one of the most widespread, and certainly one of the most adaptable, members of the dog family. It even rivals the gray wolf in terms of global distribution.

Common name Red fox

Scientific name *Vulpes vulpes*

Family	Canidae
Order	Carnivora
Size	Length head/body: 18–35.5 in (45–90 cm); tail length: 12–21.5 in (30–55 cm); height at shoulder: up to 14 in (36 cm)

Weight 7–31 lb (3–14 kg)

Key features Typical fox with long, narrow body ending in thick, brushy tail; pointed muzzle and ears; neat legs and feet; fur typically red, but varies from deep gold to dark brown, fading to white on muzzle, chest, and belly; often darker on legs; black and pale variants known

Habits Mostly nocturnal; sometimes lives in family groups, but usually hunts alone; nonbreeding males are solitary

Breeding Litters of 1–12 (usually 3–7) cubs born in spring after gestation period of 51–53 days. Weaned at 8–10 weeks; sexually mature at 10 months. May live up to 12 years in captivity, rarely more than 5 in the wild

Voice Barks, whines, yelps, screams, excited "gekkering" when playing

Diet Omnivorous; rodents and other small mammals; also insects, worms, and fruit

Habitat Diverse; includes farmland, forest, grassland, moorland, tundra, and urban areas

Distribution Europe and North America; also parts of Africa and Asia; introduced to Australia

Status Population: abundant. Persecuted as vermin; also hunted for sport

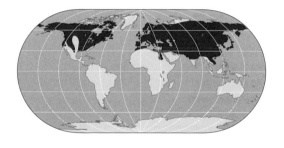

RED FOXES LIVE WILD IN NORTH America, Europe, Asia, and Africa, and have become widespread in Australia and on many islands to which they have been introduced. There is considerable variation in size and appearance throughout the range, with the largest foxes occurring in Europe. The typical red fox coat is a deep red-brown, with white on the muzzle, chest, belly, and tail tip, and black on the legs. In North America, however, there are many distinct color varieties, with up to 20 percent of red foxes being black or silver. Other varieties include so-called "cross foxes," which are basically red with a cross-shaped mark of darker fur on the shoulders. "Samson foxes" have coats that lack the normal long guard hairs and therefore look somewhat fluffier than usual.

Misrepresented

Throughout their huge geographical range foxes are loved and loathed in almost equal measure. It is difficult not to admire an animal so smart and adaptable that it is able to live almost anywhere that people can. The fox features frequently in folk tales and fairy stories, and its glorious pelt is valued as an expensive fashion accessory. However, the animal is traditionally detested by farmers and is persecuted throughout much of its range because of its predatory habits and the risk of transmitting rabies. Foxes are trapped, shot, and hunted almost everywhere they occur, and yet they still manage to thrive.

Foxes' diets and their hunting and foraging techniques vary as much as their habitat. In temperate climates in late summer many foxes exist almost entirely on sugary fruits, such as blackberries and apples. On warm, humid

ⓘ *A red fox holds an Arctic ground squirrel firmly in its jaws. The fox's sharp features and smart nature contribute to its reputation as a wanton predator. But contrary to popular opinion, foxes rarely kill more than they need.*

evenings in summer when earthworms come to the surface on open grassland and pasture, a fox can eat enough in an hour to keep it going for the whole day. At other times hunting is more intensive, and foxes will stalk, chase, and pounce on prey, including voles, rabbits, frogs, and birds. Ever the opportunist, a fox will also take advantage of roadkills and refuse. Excess food is usually stored—bones and bits of meat are buried in the ground to be dug up and eaten later, maggots and all. Rotten meat does not appear to do foxes any harm.

Chicken Runs

Foxes are messy eaters, and food remains are often scattered widely, but little is actually wasted. Contrary to popular opinion, foxes are not wanton killers and will rarely kill more than they need. Stories of foxes running amok in chicken runs and killing dozens of birds at a time have more to do with the unnatural conditions in which chickens are kept than the fox's killer instincts. In a run where chickens live at high density and have no way of escaping, a fox cannot simply make a kill and slink away to eat in peace. A flock of panicking birds causing mayhem all around sends the fox into a frenzy. As long as the chicken farmer ensures his enclosures are fox-proof, the problem does not arise.

⬆ *The red fox occurs in many distinct color forms. Above, the vivid, flame-red coloring of most high-latitude red foxes (1); the silver form (2); the "cross fox," with a cross shape on its shoulders (3).*

Communal Lifestyle

Until recently foxes were thought to be solitary animals. They are certainly territorial and tend to hunt alone. However, in the privacy of their breeding dens the story can be quite different. A single communal territory can be home to as many as six adult foxes: one dominant male (the dog-fox) and up to five vixens (females). The vixens are apparently always related, each one either a sister, mother, or daughter to the others. The male usually mates with just one of the vixens, occasionally two if the habitat is productive enough to support an extra litter.

Breeding vixens are dominant over all the others. Status within a group is often established when the vixens are very young, long before they reach breeding age, and is reinforced continually. The dominant vixen is sometimes aggressive, sometimes friendly and reassuring, but her mood can change in an instant. Her subordinates are always ready to adopt cowering, submissive postures and to make themselves scarce when she chooses to remind them who is boss. Subordinate females seem to take great pride in caring for the dominant vixen's litter and compete for the privilege of baby-sitting.

Red Fox Cubs

Baby foxes are born in litters of one to 12, the average number varying according to the quality of habitat. The cubs are born blind but furry, and each weighs between 2 and 6 ounces (57 and 170 g). To begin with, their fur is dark chocolate brown, and their eyes, which open after two weeks, are blue. By the time the youngsters are one month old and ready to leave the safety of the breeding den for the first time, they have already begun to look more like foxes. Their fur lightens, their eyes turn brown, and their muzzles are longer and more pointed.

⊖ Urban foxes make a good living feeding on refuse and bird-feeder leftovers, and by killing rats, pigeons, and other town-dwelling wildlife. Often the pickings are so rich that foxes in towns and cities live at much higher densities than they ever manage in the countryside.

A young fox's first taste of meat is usually in the form of partially digested scraps coughed up by its mother. Later the cub's milk teeth drop out and are replaced by the adult dentition. The jaws and teeth are strengthened by chewing on bones, sticks, and other objects, and the cub's coordination and hunting skills are developed by hour on hour of boisterous play with its siblings. Adult foxes retain a playful streak, and games can involve the whole family in a noisy rough-and-tumble.

Habitat Requirements

Young females may stay with the family group, but males always disperse, traveling about 30 miles (48 km) or sometimes farther, to establish their own territory. The size of a fox's territory depends on the quality of the habitat and especially on the availability of food. Ideal fox habitat has a selection of different habitat types: Areas of woodland and pasture crisscrossed with hedgerows and the odd garden are ideal. Sometimes a fox can find all it needs in a territory of about 25 acres (10 ha). In less hospitable habitats, such as the Canadian tundra, a fox may require a hundred times as much space to supply its needs. Territories are diligently marked with urine and droppings.

⊕ A female red fox with a cub. The cubs are ready to leave the safety of the breeding den at about one month. They can fend for themselves at six months and breed at 10 months.

Deadly Virus

One of the most serious and widespread threats to foxes other than human persecution is the rabies virus. Rabies is found in much of the world's fox population, except in Britain, whose strict quarantine laws have kept the disease from becoming established. In continental Europe rabies has been largely eliminated from the fox population in several countries by using special vaccines distributed in baits, which the wild foxes eat. Vaccination programs reduce the threat of rabies because the disease dies out when there are enough foxes in the population that are immune to it. The method is expensive but humane and can be very effective as long as enough baits are distributed.

Swift Fox

Vulpes velox

The attractive swift fox is small in stature and active mainly at night. Unusually for a fox, it is also bold and curious, and therefore easy to snare. In the past swift foxes were ruthlessly hunted for their highly prized fur.

Common name Swift fox (kit fox)

Scientific name *Vulpes velox*

Family Canidae

Order Carnivora

Size Length head/body: 15–21 in (38–52 cm); tail length: 9–14 in (22–35 cm); height at shoulder: 12 in (30 cm)

Weight 4–7 lb (2–3 kg)

Key features Similar to red fox; winter coat grayish-beige with pale undersides and rich orange-brown on legs, tail, and flanks; summer coat shorter and darker; bushy tail tipped with black and slightly shorter than in other foxes

Habits Active at night; social, bold, and tame

Breeding Three to 6 young born in spring after gestation period of 50–60 days. Weaned at 6–7 weeks; sexually mature at 10 months. May live up to 14 years in captivity, usually fewer than 6 in the wild

Voice Quiet yelps and barks

Diet Small mammals, especially rabbits, pikas, and rodents; also birds, lizards, amphibians, insects, and occasionally plant material

Habitat Prairie and grassland

Distribution Scattered populations across central plains of the U.S. and Canada

Status Population: low thousands; IUCN Endangered (northern subspecies), elsewhere Lower Risk: conservation dependent; CITES I (northern subspecies)

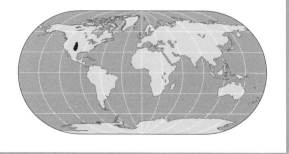

THE SMALLEST OF NORTH AMERICA'S nine species of wild dog, the swift fox is about the size of a domestic cat. Swift foxes are generally nocturnal (active at night) and spend the day holed up in burrows, which they either dig for themselves or modify from the excavations of other prairie mammals. The burrows are fairly large and may have several entrances. During the spring and summer a burrow may contain as many as nine or 10 foxes, including a breeding pair and their offspring of the year. Occasionally a second adult female may also be present in the burrow. It takes the efforts of at least two adults to rear a litter of pups, which may number three to six, but the pair bond does not necessarily last from year to year.

Specialized Diet

A typical family will occupy a home range of about 12 square miles (31 sq. km), which is unusually large for a fox and reflects the species' more specialized food requirements. While most other foxes will eat almost anything, swift foxes are rather choosy about what they consume. Their preferred prey consists of gophers, pikas, and rabbits, although they will eat other animals and even grass and berries if there is no alternative. Swift foxes do particularly well in places where prey animals are plentiful, and their impressive acceleration is invaluable in catching the fleet-footed creatures they pursue. Over even terrain swift foxes can easily reach speeds of 30 miles per hour (48 km/h).

Swift foxes once roamed all over the central plains of North America, from Texas in the south to Saskatchewan and Alberta in the north. Today, however, the animals occupy a mere 10 percent of that area, living in small, scattered populations with larger numbers at the center of their distributional range in Colorado and Wyoming. The relentless spread of agriculture across the prairies ruined much of the habitat of the swift fox. Plowing destroyed their burrows and those of their prey, and the planting of crops has changed the nature of the prairies forever.

Mistaken Identity

During the mid- to late 19th century there was a vigorous effort, supported by the United States government, to eradicate wolves and coyotes from the entire continent. Although swift foxes were not considered to be vermin, they could not be prevented from eating the poisoned baits put out to kill coyotes. Another problem was that they were (and still are) often mistaken for young coyotes, and many have been shot as a result of misidentification.

Unusually for foxes, the swift fox is bold and curious and therefore relatively easy to snare, shoot, and poison. Moreover, swift fox fur could fetch a good price, so there was every reason for the killing to continue. In the United States the foxes were wiped out of all but the southern part of their range by 1920. However, in the mid-20th century the species began to recover, spreading north once more into Wyoming, Nebraska, Oklahoma, Montana, and the Dakotas. But north of the border there was no such recovery, and by 1978 there were no swift foxes left in Canada.

It has taken an expensive and long-running program of reintroduction to restore a small population of swift foxes to southern Alberta and Saskatchewan. The northern populations are different enough from those in the south to be considered a separate subspecies, known as *Vulpes velox hebes*. They are darker than the southern foxes and have a broader muzzle.

⊖ *For nearly a century the diminutive swift fox suffered persecution through mistaken identity and trapping for its valuable fur.*

Common name Arctic fox

Scientific name Vulpes (Alopex) lagopus

Family Canidae

Order Carnivora

Size Length head/body: 18–27 in (46–68 cm); tail length: 12 in (30 cm); height at shoulder: 11 in (28 cm)

Weight 3–20 lb (1.4–9 kg)

Key features Stout-looking fox with short legs, long, bushy tail, small, rounded ears, and a thick, woolly coat; fur pure white in winter in high arctic animals; fur extends to soles of feet

Habits Social; sometimes migratory; active at any time of day; does not hibernate

Breeding Litters of 6–12 (occasionally as many as 25) pups born in early summer after gestation period of 49–57 days. Weaned at 2–4 weeks; sexually mature at 10 months. May live up to 16 years in captivity, many fewer in the wild

Voice Barks, whines, screams, and hisses

Diet Mainly carnivorous; prey includes seals, rodents (especially lemmings), seabirds, fish, invertebrates such as crabs, mollusks, and insects, and carrion; scavenges from kills made by other arctic predators; occasionally plant material

Habitat Arctic and northern alpine tundra, boreal forest, ice cap, and even sea ice

Distribution Arctic regions of Canada, Alaska, Greenland, Iceland, Finland, Sweden, Norway, and Russia

Status Population: abundant. Generally common, although range and population size have declined recently. Protected in Norway, Sweden, and Finland; hunted for fur and as vermin elsewhere

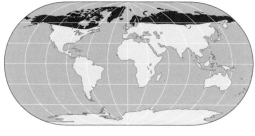

Arctic Fox

Vulpes (Alopex) lagopus

The Arctic fox has a number of adaptations in order to survive in extremely cold conditions. Nevertheless, life for the species can be very difficult, with many deprivations to endure.

AS ITS NAME IMPLIES, THE RANGE of the hardy Arctic fox extends well beyond the Arctic Circle and farther north than any other member of the dog family. Arctic foxes have been recorded at latitudes as high as 88°N, only 150 miles (240 km) from the North Pole itself. But the foxes are only visitors to these frozen wastelands, and normally live farther south, especially in areas of coastal tundra in Canada, Greenland, Iceland, and northern Europe.

The southern edge of the Arctic fox's geographical range seems to be where the northern range of the red fox ends. Red foxes can make a living almost anywhere there is enough food to support them, but they are defeated by extreme cold. The Arctic fox here comes into its own, being able to tolerate temperatures that sometimes plummet to -58°F (-50°C). Captive individuals have been able to survive at -112°F (-80°C) under experimental conditions. Few other animals can tolerate such extreme cold.

Cold Weather Design

Both red and Arctic foxes have basically the same body shape, but the Arctic fox has relatively shorter legs and ears and a smaller muzzle. The Arctic fox's more compact shape is designed to lose less heat than the red fox's long and lean frame, which is built for speed and agility. Only the tail of the Arctic fox is very long, which allows it to be used as a kind of blanket to cover the fox's face while it sleeps. The most important adaptation to the cold is the Arctic fox's luxuriant fur. Said to be the warmest fur in the animal kingdom, it is fine, long, and fluffy. It is also incredibly dense, and in winter it grows to be three times as deep as

There are two color varieties of Arctic fox, known as white and blue. However, the white variety is only white in the winter months of October through April. Over the summer the fur usually turns a grayish-brown. White-furred foxes dominate the populations in Canada. In Greenland about half the foxes are white in winter, but in Iceland almost all of them are blue. Blue foxes are actually a steely gray color, which is darker in summer than in winter.

In addition to fur and body shape the Arctic fox has made a number of physiological adaptations to the cold. The fleshy parts of its paws are well supplied with blood vessels. Here, an extensive network of fine capillaries brings warm blood to the feet and toes, helping prevent frostbite. After passing through the feet, the cooled blood travels back up the leg past numerous other vessels carrying warm blood from the heart. The returning blood is in this way rewarmed before it enters the rest of the body to avoid it causing a drop in the fox's core body temperature.

Conserving Energy

In times of abundant food fat is accumulated under the skin, providing both insulation against the cold (like the blubber on a seal) and a reserve of energy for when food is scarce. In winter almost half the Arctic fox's body weight is fat. In especially hard times, when an Arctic fox has not eaten for many days, it is able to slow down its metabolism to about half the usual rate to save energy. The fox then has to be much less active than normal: It may even lay up in a snow hole for a while, but it does not actually hibernate. It stays fully alert and can spring into life as soon as a feeding opportunity arises.

Snow holes provide temporary shelter for wandering foxes, but for breeding purposes the animals require something more substantial. Arctic foxes build extensive dens,

⊕ An Arctic fox displaying pure white winter coloration. The fur of the Arctic fox is said to be the warmest of all animals: The hairs of the coat are hollow to provide extra insulation in subzero temperatures.

in summer. Like the hairs in a polar bear's coat, those of the Arctic fox are hollow. Each individual hair therefore contains air, which helps provide extra insulation. Fur even grows on the soles of the fox's feet, protecting them from the chill of ice and snow. They also help the fox get a better grip in slippery, icy conditions. These furry feet are the reason for the animal's scientific name of *lagopus*, which literally means "rabbit-footed."

often in the base of a cliff or in a mound of earth and stones. Some of these dens have been in more or less continuous use for hundreds of years by generations of foxes. A den typically has several entrances, usually between four and a dozen, but sometimes up to 100. Long-established dens become quite a feature of the landscape, with taller vegetation growing around the entrances compared with elsewhere on the tundra. That is because of the extra nutrients from fox droppings and waste food that encourage the plants to grow.

Slim Chance of Survival

Arctic foxes are social animals, but groups are quite small: typically one breeding pair and their young of the year, plus a helper female (one of the previous year's offspring). Adult foxes mate for life and it takes all their efforts to raise a litter in what can be difficult conditions, even in summer. Litters are large, sometimes over 20 pups, but usually six to 12. The assistance of the young helper female means that two adults can hunt while a third stays at the den to baby-sit. Even so, the chances of any one youngster living longer than a few months are low and many die long before their first birthday. If they survive until the fall, the young foxes disperse to make their own way in life.

During the winter some foxes remain near the breeding den (especially if food is plentiful), but others undertake some of the most extraordinary journeys known in the animal kingdom. Sometimes they travel hundreds of miles from land, far out over the frozen sea. Foxes are not averse to swimming where necessary and can travel many miles by hitching a ride on an ice floe. In the winter the only food for foxes is whatever they can scavenge from the kills of polar bears. Beggars cannot be choosers, and Arctic foxes will eat anything from rotten meat to feces. Farther south and in summer their menu is more varied and includes birds, berries, and small mammals. For many fox populations lemmings form the main diet and at times are staggeringly abundant. However, every few years the lemming population crashes, and the foxes starve or are forced to search for alternative food, sometimes venturing hundreds of miles outside their normal range and far south of the snow line.

⊖ Arctic fox pups at their den. Litters are large (anything up to 25 pups), and parents are often assisted in baby-sitting duties by a female helper. Even so, the chances of a pup's survival are low.

⊖ A barking Arctic fox displays its short summer coat. Over the summer months of May through September the fur usually turns a grayish-brown color. The coat can be up to three times as dense in winter as in summer.

The Fox Fur Trade

Arctic foxes have been hunted for their fur for hundreds of years. They are trapped in snares or shot almost everywhere except Scandinavia, where the population is very small and threatened with extinction. At the height of the fur trade foxes were extensively farmed in places like Alaska. Blue fox fur is considered more valuable by the fur trade, and the blue foxes living in Alaska and on the Aleutian Islands are almost all descendants of animals that escaped from fur farms. A good blue pelt can fetch about $300; but fashions change, and the demand is not as great as it once was. White foxes are hunted, too, but their fur is less valuable—in fact, it is sometimes dyed blue-gray in order to fetch a better price.

An Arctic fox caught in a trap. Foxes have been hunted for hundreds of years.

Common name Polar bear

Scientific name *Ursus maritimus*

Family Ursidae

Order Carnivora

Size Length head/body: 6.6–8.2 ft (2–2.5 m); tail length: 3–5 in (7–13 cm); height at shoulder: up to 5.2 ft (1.6 m)

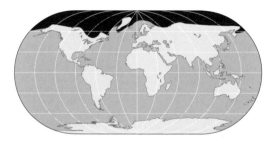

Weight Male 660–1,760 lb (300–800 kg); female 330–660 lb (50–300 kg)

Key features Huge bear with thick, off-white coat; head relatively small; feet large and furry

Habits Solitary; migratory and partially nomadic; pregnant females hibernate in winter; excellent swimmer

Breeding Litters of 1–4 tiny cubs born in midwinter after gestation period of 195–265 days (includes variable period of delayed implantation). Weaned from 6 months; sexually mature at 5–6 years. May live up to 45 years in captivity, 30 in the wild

Voice Grunts and growls

Diet Carnivorous: mainly seals but occasionally other animals such as reindeer; also fish, seabirds, carrion, and plant material in summer

Habitat Sea ice, ice cap, and tundra; equally at home in water and on land

Distribution Arctic Circle; parts of Canada, Alaska, Russia, Scandinavia, and Greenland

Status Population: 20,000–30,000; IUCN Lower Risk: conservation dependent; CITES II. Main threat is from human exploitation of Arctic habitats

Polar Bear

Ursus maritimus

The polar bear is the world's largest land carnivore and is superbly adapted to life in one of the harshest regions on earth.

POLAR BEARS AND BROWN BEARS are more closely related than their appearance and different lifestyles suggest. Until about 100,000 years ago they were the same species, and even today individuals in captivity are able to interbreed. The special features that allow polar bears to survive life in and out of the water in one of the bleakest, most inhospitable parts of the world are all fairly recent adaptations, providing a good example of how evolution can proceed quickly under extreme conditions.

Cold Weather Protection

The most striking polar bear characteristic is, of course, its color. But there is more to the coat than meets the eye. Not only are the hairs very long, trapping a deep layer of warm air against the skin, but (under a microscope) the individual hairs can be seen to be hollow. Each has air spaces running along its length, which help make the coat extra warm because air trapped inside the hairs improves the insulation effect. It is the hollowness of the hairs and the lack of pigment that makes the fur appear white. The dense coat is also surprisingly light. Late in the season, before the fur is molted, it begins to look rather yellow, owing to a combination of accumulated dirt and the oxidizing effect of sunlight. Zoo polar bears sometimes get algae from their pool into the coat hairs, turning them temporarily green, but this does not happen in the wild!

The other obvious feature of polar bears is their size. Fully grown males are the world's largest terrestrial predators, measuring about 8 feet (2.5 m) long on all fours and weighing as much as 10 large men. Females are less than half this size, but still number among the world's most powerful animals. Large body size

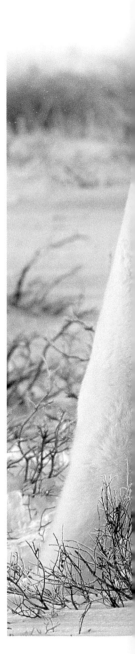

is another adaptation to the cold, because larger animals are more efficient at preventing loss of body heat. Being big also allows polar bears to tackle large prey—for much of the year seals are the only other animals around, and a polar bear can scoop one out of the water using just a single paw. Another special adaptation is the huge furry feet, which help spread the polar bear's weight so effectively that a bear weighing half a ton (508 kg) can walk carefully across ice too thin to support a human. The soles are also furry, protecting the pads from frostbite and giving the bear extra grip on the ice: They also help reduce the tendency to sink into soft snow. The bears are

⊕ Polar bears test their strength in bouts of play wrestling. Fully grown males are the world's largest terrestrial predators and can weigh as much as 10 large men.

nimble for their size and can gallop at speeds of up to 30 miles per hour (50 km/h) for relatively short periods.

Long-Distance Traveler

Polar bears have never actually been recorded at the North Pole. Here the sea ice is thick and continuous, with no access to open water for the bears or for the seals on which they feed. However, they do occur almost everywhere else within the Arctic Circle, concentrating their activity around the thin, cracked edges of the pack ice where seals haul out. In winter, when the sea ice reaches its maximum extent, the bears venture as far south as Newfoundland, southern Greenland, and the Bering Sea.

Polar bears wander widely, but they are not true nomads as was once believed. Recent studies show that bears from different parts of the Arctic form distinct populations, with individual bears using ranges of up to 200,000 square miles (500,000 sq. km) over the course of a few years. There is a resident population of polar bears around the Hudson and James Bays,

⊕ A female polar bear with her cubs. Polar bear cubs stay with their mother for about two and a half years, only leaving when she is ready to breed again.

Delayed Implantation

Most polar bears are solitary and wander over vast areas. Males and females rarely meet, so they are ready to mate whenever the opportunity occurs between March and June. Whatever the time of mating, the cubs are nearly always born in midwinter. It is the best time of year for births because it allows the maximum period for growth and development after the babies have left the den. A polar bear's pregnancy can therefore be anything from six and a half to almost nine months. Soon after fertilization of the mother's eggs the tiny embryos go into a state of suspended animation. It is the fact that the embryos do not begin to develop immediately that makes the variable gestation period possible. Pregnancy and rearing cubs over the winter put a huge strain on the female's body and can be fatal if she is not in good health. Delaying the development of the embryos until the female has put on enough weight to survive the pregnancy and provide milk for the cubs through the winter guards against starvation of the entire family. If the female is not in top condition by the late fall, the embryos are spontaneously aborted.

members of which do not need to travel so far. They spend their summers on land, venturing up to 120 miles (200 km) inland, and move out onto the vast expanse of ice when the bays freeze over in winter.

Smash and Grab

Ringed seals are the most important prey species, and polar bears show considerable flexibility in the techniques used to hunt them. In late spring female ringed seals give birth to their young in well-hidden dens. The dens have openings to the sea below but are invisible from above, being roofed over with snow. However, polar bears have an acute sense of smell and can detect the pups lying quietly below. They break into the den using brute force, rearing up on their hind legs and pounding the roof with their front feet. They then seize the seal pup inside. Hunting adult seals, on the other hand, is all about stealth and patience. Bears wait silently by a breathing hole for a seal to emerge, then grab it and heave the animal onto the ice. Sometimes the bears sneak up on a seal resting on the ice, using snow ridges and ice blocks as cover. They creep forward in a low crouch, keeping still every time the seal looks around. Not every hunt is successful, but the bears often kill enough to feed not only themselves but an entourage of scavenging Arctic foxes as well.

Varied Diet

Individual bears have distinctly different hunting techniques, which they develop according to their own experience. Other items that may appear on the polar bear's menu include harp and bearded seals, young beluga whales, walrus, reindeer, fish, seabirds, dead animals, and occasionally plant material. Bears arriving on land in the summer may spend hours browsing on leaves and berries, which, although not especially nutritious, contain some vitamins and minerals otherwise completely lacking in the bear's diet. For many bears summer is a time of hunger because the lack of sea ice means they cannot hunt seals. The Hudson Bay bears may go for months without eating, living only on their fat reserves and staying as inactive as possible to save energy and to avoid overheating in the weak sunshine.

Breeding Dens

Most polar bears remain active throughout the winter, only seeking shelter in temporary snow holes during the worst storms. They do not normally need to hibernate because there is no shortage of food at this time of year. Pregnant females, however, build substantial dens in which to spend the winter. The dens, which are dug into a bank of snow, usually consist of a tunnel up to 10 feet (3 m) long and a large oval chamber. Some are rather more elaborate and may have several interconnected rooms. The female sleeps in the den throughout the winter,

The Sea Bear

The polar bear could just as correctly be called the sea bear (indeed, its scientific name means precisely that). It is a superb swimmer and is just as comfortable in the icy water of the Arctic Ocean as on land or pack ice. Polar bears can float effortlessly in seawater and do not sink even when dead. The hollow hairs in their coat are much more buoyant than normal fur. The fur is also slightly greasy and repels water. After a swim the bear only needs one quick shake to remove most of the moisture from its coat, so there is little danger of ice forming in the fur. The toes of the bear's enormous paddle-shaped feet are slightly webbed, making them more effective for swimming. The bear's neck is long, and it swims with its head held high above the water so that it has a good view over the waves.

Polar bears can swim for hours, using a steady dog paddle. They have even been known to swim up to 40 miles (65 km) across open water. They can dive under ice and climb out through seal breathing holes or leap 7 feet (2 m) onto ice cliffs. Hitching a ride on a passing ice floe is a favorite way of getting around, and the bears seem quite happy to plunge in and out of the cold water dozens of times a day.

⊙ **Polar bears are excellent swimmers and can paddle for hours at a time.**

during which time the cubs are born. They are very small and need protecting from the harsh climate for the first few months of life. The newborn young make their own way to their mother's teats, and she suckles them without appearing to wake up. This long sleep is not true hibernation because although the female's heart rate and breathing slow down, her body temperature only drops by a few degrees. As a result, the den remains cozy, and she can wake up quickly if need be. By the time spring comes, the cubs have increased in weight from just over 1 pound (500 g) to between 25 and 30 pounds (11 and 14 kg) apiece. The mother is half-starved, having used up most of her fat to produce milk. Her first priority is to find food, but that is not easy with up to three lively cubs romping by her side.

Bear Attacks

Polar bears are aggressive. They can and do kill humans; but since little of their range is populated, the number of fatalities is low. People who live and work within the polar bear's range are generally well informed when it comes to bears, and visitors are given plenty of advice on how to avoid danger. Bear attacks are most frequent in the Hudson Bay area, especially around the town of Churchill, where several people have been attacked in the last 40 years. The bears pass by the town on their regular migrations and are attracted to the municipal waste dumps where they are liable to attack anyone who disturbs them.

⊝ *Polar bears wander widely, with individuals using ranges of up to 200,000 square miles (500,000 sq. km) over the course of a few years.*

Polar Bears and Humans

Polar bears have been known to the Inuit people from the time they settled in the North American Arctic about 4,000 years ago. The bears figure prominently in native folklore and spirituality. They were traditionally hunted for meat, fur, and other body parts. More recently polar bears were also hunted commercially, but the practice ceased in 1976 as the result of an agreement between the five "Polar Bear Nations"—the United States, Canada, Norway, Russia, and Denmark. Conservation laws now include controls on commercial hunting: Most of the bears hunted today are killed as part of the traditional Inuit hunt. However, hunting is not the only threat, and polar bears currently face problems associated with pollution and the exploitation of the Arctic for mining and oil extraction.

Common name American black bear

Scientific name *Ursus americanus*

Family Ursidae

Order Carnivora

Size Length head/body: 4.9–5.9 ft (1.5–1.8 m); tail length: 4.5 in (12 cm); height at shoulder: up to 36 in (91 cm)

Weight Male 250–600 lb (113–272 kg); female 200–310 lb (91–141 kg)

Key features Large bear with thick, but not shaggy coat; fur can be variety of colors, but usually brown or black; muzzle less furry than rest of face

Habits Solitary; most active at night; swims and climbs well; hibernates over winter

Breeding Litters of 1–5 (usually 2 or 3) cubs born after gestation period of 220 days (including about 150 days delayed implantation). Weaned at 6–8 months; females sexually mature at 4–5 years, males at 5–6 years. May live up to 31 years in captivity, 26 in the wild

Voice Various grunts, rumbling growls, and woofing sounds; cubs give high-pitched howls

Diet Mostly plant material, including fruit, nuts, grass, bark, and roots; fish; invertebrates such as insects and their larvae and worms; also honey, other mammals, and carrion

Habitat Forest and scrub; occasionally open spaces

Distribution Canada, Alaska, and U.S. south to Mexico

Status Population: 400,000–500,000; CITES II. Still common, but population now reduced due to hunting, persecution, and habitat loss

American Black Bear

Ursus americanus

Think of the American black bear, and the image of the cartoon character "Yogi Bear" may well come to mind. In fact, the black bear is very similar to the adaptable and opportunistic forager portrayed in the TV show.

THE AMERICAN BLACK BEAR HAS always been widespread and remains common over much of its huge geographical range. It is a typical bear, and its enterprising and opportunistic behavior has made it the inspiration for myths, folk tales and modern stories, films, and cartoons. Even so, the details of its biology have only recently become well understood. Before the 1960s and 1970s zoologists lacked the necessary skills and technology to make safe, nondisruptive, long-term observations of these shy but powerful animals in the wild. Since that time live trapping, radio tracking, and other up-to-date techniques have provided much information about the bear's daily life and behavior patterns.

Color Variations

American black bears can be a variety of colors ranging from white to black and including reddish and chocolate brown, bluish-black, and dark blond. However, dark-brown or black individuals are by far the most common. There are geographical trends in color variation, with most nonblack bears occurring in the southwest in California and Mexico. The white form (not to be confused with the polar bear, which is another species entirely) is rare, but most white bears come from the Pacific coast of southwestern Canada.

Black bears are not particularly aggressive toward people; but like all large animals, they can be unpredictable and dangerous, especially if they are injured, frightened, or provoked. Black bears generally avoid confrontations with humans, but their opportunistic foraging habits can bring them very close to areas of human

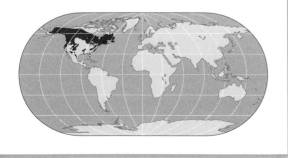

⬆ *An adult black bear rests on a fallen log. Black bear habitat is usually forest or scrubland, but bears will occasionally venture into open spaces and even the fringes of towns.*

activity and thus into potentially tragic situations. Fatal black bear attacks are rare, but they always receive high-profile publicity.

Black bears normally live alone, except for courting pairs, mothers with cubs, and occasional gatherings around a plentiful food resource, such as a waste dump. Each has a separate territory; but home ranges can overlap, so that in areas of prime habitat, such as Washington State's Long Island and parts of California, they can live at densities of two or three bears per square mile (about one per sq. km). Male bears are more territorial than females, and the overlaps between their ranges are small or nonexistent. However, the size and shape of a male's home range are usually determined by the ranges of the local females—the male occupies a space that gives him access to as many potential mates as possible.

While the bears in a local area usually keep to themselves as much as possible, there is often a loosely structured social hierarchy, which comes into play when the bears meet.

⊕ *Coat color ranges from black to white, depending on geographical location. Bears from southeastern Alaska (above) have a bluish-gray coloration.*

"Smarter than the Average Bear"

So goes the catchphrase of the famous cartoon character Yogi Bear, and it is a true description of the cleverness of black bears. The species will eat just about anything it can lay its paws on; but like people, black bears are especially partial to high-energy foods and those with a high fat and protein content. Yogi Bear and his sidekick Booboo are fictional of course, but their endless quest for unguarded picnic hampers is not far short of reality. National parks throughout the black bear's range have strict rules on the safe storage and disposal of food. Backpackers are advised never to keep food in their tents to reduce the risk of nighttime raids by clever bears who have learned where to get an easy meal. Official campgrounds usually provide lockable metal boxes in which food can be kept out of reach. Wilderness campers are advised to use hanging larders, since even smart bears find them difficult to break into. Park bears have become used to people, cars, and roads, and are intelligent enough to recognize sealed soft drink cans and other unnatural-looking objects as food.

A picnic hamper is never safe with Yogi and Booboo around!

For example, males competing for a female will size each other up with aggressive posturing, rearing up on two legs, and wrestling with one another. An inferior bear will back down, while two closely matched bears may come to serious blows. The bear that emerges victorious from the bout will probably retain his dominance next time the two meet, so avoiding the need for more violence.

Moving In

Good black bear country is rugged, with plenty of tree cover. Historically black bears were probably discouraged from venturing far onto the Canadian tundra at the north of their range by the presence of brown/grizzly bears. Grizzlies not only present stiff competition for food and shelter, but they will occasionally kill small black bears. However, where the tundra grizzlies have declined (due to hunting and persecution), it seems that the black bear is only too happy to move into the vacated territory to forage. One of the most

⊕ *A cinnamon-colored mother (sow) with her cubs. Up to five cubs are born in January or February. They stay in their underground den, suckling from their mother, until quite late in the spring.*

⊕ *The opportunistic foraging habits of bears can often bring them into close contact with humans, sometimes with tragic consequences. Here, an American black bear scavenges at a waste dump in Canada.*

important requirements of bear country is enough suitable hibernation sites. Black bears hibernate because there is not enough food available during the winter to sustain them in normal activity. Many of the bear's natural food sources are highly seasonal, with fruit, berries, and nuts all peaking in late summer.

During this time of plenty the bears gorge themselves, becoming fat and lethargic. By mid-fall, when most of the food is gone, they stop eating and seek out a secure den in which to spend the winter. It might be a cave, a hollowed-out log, or the space under a fallen tree. Some dens are used every year by the same or different individuals.

Winter Slumber

Once asleep, the black bear's core body temperature drops four to seven degrees to between 93.2 and 87.8°F (34 and 31°C). Its breathing and heart rate slow right down until its metabolism is just ticking over using the bare minimum of energy. The bear will stay in that torpid state as long as the cold weather lasts, but will rouse during short periods of warm weather, sometimes even emerging from the den for a day or two. Bears that live in the north hibernate for longer than those in the south, and the winter sleep can last anything from 75 to 130 days. When they emerge from

their den, the bears will selectively forage for the richest food in order to regain the weight they lost over the winter months of hibernation.

Females that mated the previous summer may give birth while they hibernate. Like many other carnivores, American black bear embryos undergo a period of suspended development soon after they are conceived. They do not implant into the mother's uterus until her body is in prime condition and she has put on enough weight to be able to support herself and her developing cubs during the winter. The mother bear may not get the chance to eat again until quite late in the spring when the cubs will be two or three months old.

The cubs are born in January or February and are virtually naked when they first appear. They weigh only 8 ounces (230 g). They suckle from their sleeping mother's teats, putting on weight and becoming livelier almost by the day. By the time spring arrives and the family emerges from the den, the cubs are fully furred bundles of energy. They continue to suckle for a further four to six months and are gradually weaned onto solid food, which their mother teaches them to find. The young bears will spend the whole summer and the following winter with their mother. They will usually disperse at about 18 months of age, leaving her free to have another family.

Common name Brown bear (grizzly bear, big brown bear)

Scientific name *Ursus arctos*

Family Ursidae

Order Carnivora

Size Length head/body: 5.5–9.3 ft (1.7–2.8 m); tail length: 2.5–8 in (6–20 cm); height at shoulder: 35–60 in (90–150 cm). Male bigger than female

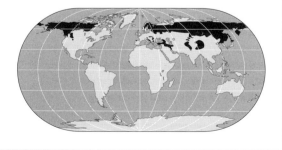

Weight 132–1,750 lb (60–800 kg)

Key features Medium to large bear with shaggy, light-brown to black fur, often grizzled (grayish) on back and shoulders; narrow snout; broad face

Habits Solitary; nonterritorial; hibernates over winter

Breeding Litters of 1–4 (usually 2) cubs born January–March after gestation period of 180–266 days. Weaned at 5 months; sexually mature at 4–6 years. May live up to 40 years in captivity, 25 in the wild

Voice Various grunts and growls

Diet Mostly plant material, including grass, roots, and fungi; also invertebrates such as worms and insects and their larvae; fish and carrion

Habitat Varied; tundra, open plains, alpine grassland, forests, and wooded areas

Distribution Western Canada, Alaska, and northwestern U.S.; northern Asia south of Arctic Circle; Scandinavia, eastern Europe, and Middle East; Pyrenees, Alps, and Abruzzi Mountains

Status Population: 220,000; CITES I (several Eurasian subspecies); CITES II (North American subspecies). Declining, but now more stable

Brown/Grizzly Bear

Ursus arctos

This highly successful and widespread bear ranges widely across the Northern Hemisphere. The largest brown bears occur in the United States and are often referred to as grizzly bears.

AMONG THE WORLD'S MOST FEARED and admired carnivores, the brown bear (known as the grizzly bear in parts of North America) is also one of the largest and most prevalent. At its most widespread its range covered most of North America, Europe, and Asia. It still occurs widely in all three continents; but its distribution is now more patchy, and some smaller populations are seriously threatened with extinction. Because of its size and strength the brown bear has been a feared neighbor for country people over many centuries.

Regional Variations

Brown bears from different geographical areas vary greatly in appearance and behavior. Most are some shade of brown, but off-white and almost black individuals are known. The largest bears are found on the Pacific coast of Alaska, specifically on Kodiak and Admiralty Islands. Males here reach almost 1,750 pounds (800 kg) in weight, rivaling the largest polar bears in size.

In dramatic contrast to the Kodiak giants, brown bears living just a few hundred miles away in the Yukon rarely exceed 330 pounds (150 kg), and those in southern Europe are often under 150 pounds (70 kg). Such huge variation in size might suggest that there is more than one species of brown bear, but in fact the difference probably has more to do with diet than genetics. Bears keep on growing well into adult life, and the rate of growth is highly dependent on the quality of food. Bears that manage to consume a high-protein diet grow much bigger than those forced to survive

on berries and grass. Kodiak bears and those living on the other side of the Bering Strait in Kamchatka benefit from the annual run of Pacific salmon, which swim upriver in their millions to spawn. For several weeks every summer the bears can gorge themselves on highly nutritious fish, loafing around on the riverbanks between meals and expending little energy. In other parts of the world bears have to survive on much more limited rations, which may also take a good deal of energy to find.

Seasonal Produce

Most brown bears eat more plant material than anything else, carefully selecting the most succulent and nutritious of the season's grasses, fruit, nuts, and fungi. They tend to avoid old-growth vegetation because it is much harder to digest, especially since their gut is basically that of a meat-eater. Brown bears kill and eat other animals—from mice to bison and other bears—as and when the opportunity arises. However, in most parts of their range predatory behavior is rarely planned in advance.

Whether hunting or foraging, the most important bear sense is smell. Compared with its huge black nose, the brown bear's eyes and ears are small, reflecting its relatively poor eyesight and hearing. Large prey animals are usually chased over a short distance at speeds of up to 30 miles per hour (50 km/h), then killed with a mighty blow from the front paws. Large grizzlies are immensely strong and can kill animals as big as horses and cattle, dragging them 100 yards (90 m) or more to feed in a safe place. Attacks on humans are rare, but always well publicized, and will probably become more frequent as the

↩ *A brown or grizzly bear from the Rocky Mountains. Brown bears vary in size, with the biggest males weighing up to 1,750 pounds (800 kg). Although somewhat lumbering in appearance, even the largest bears can run with surprising speed and agility.*

recreational use of wilderness areas increases. Most attacks involve some kind of provocation; others may be accidental, for example, when a dominant male mistakes a human for a subordinate bear. Mothers with cubs are especially aggressive, but where possible, even they prefer to usher their family away to safety rather than confront a human being. The motivation for an attack appears not to be food, since bears rarely eat their human victims.

Light Sleeper

All brown bears are capable of hibernating, and most do so for between three and seven months of the year. Hibernation is a response to poor weather and lack of food. However, for some southern brown bears conditions never get bad enough to make such a winter retreat worthwhile. Even in northern areas brown bears do not hibernate as deeply as American black bears, and they rouse quickly in response to warmer weather or disturbance of the den. Like American black and polar bears, pregnant female brown bears usually give birth in midwinter. The development of the cubs will only proceed that far if the mother is in a fit condition to rear them. Brown bears reproduce slowly: A female rarely breeds more than once in every three or four years.

Young males disperse up to 60 miles (100 km) from their birthplace. They spend the next few years waiting for the opportunity to replace a resident male or to steal a mating with a receptive female. Young females stay closer to home, often continuing to associate with each other and their mother long after the next batch of cubs is born. Such close family ties make brown bears almost sociable. In parts of the western United States large groups of bears may gather at a food source. Although they interact peacefully most of the time, there is a strict hierarchy, which may be maintained with aggressive displays and fighting.

⊖ *A brown bear fishing for salmon in an Alaskan river. Every summer the bears gorge themselves on the nutritious fish as they swim upriver to spawn.*

The Bear Trade

Bears are popular zoo animals. Some so-called "dancing bears" used to be taken from one town to another to give public performances. However, the decline of brown and black bears was due almost entirely to hunting. In North America the bears were hunted for their fur and to protect livestock. Today hunting is strictly regulated, and bears are treated as game animals rather than a commercial resource. Trading in bear body parts is also restricted by treaty. The threat to bear populations varies from place to place, which is why some animals are officially registered as needing urgent protection, while others are considered to be less at risk.

The most serious threat to bears comes from the Asian medicine trade. Paws, bones, and internal organs are all highly valued, especially the gallbladder, which can fetch over $1,000. Because bear bile is one of the few Asian medicines that may have at least some basis in science, some countries permit the farming of Asiatic black and brown bears. Farmed animals have plastic tubes surgically implanted into their gallbladders so that the bile can be drained off and used without killing the bear. This highly controversial activity is argued by some to reduce the pressure on wild populations, but it also helps perpetuate the use of bear products in the treatment of conditions for which there are a number of effective man-made drugs.

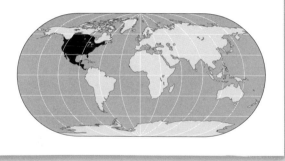

Common name
Common raccoon

Scientific name
Procyon lotor

Family Procyonidae

Order Carnivora

Size Length head/body: 18–27 in (45–68 cm); tail length: 8–12 in (20–30 cm); height at shoulder: about 10–12 in (25–30 cm). Male about 25% larger than female

Weight 11–18 lb (5–8 kg), but sometimes up to 33 lb (15 kg)

Key features Black "bandit" face mask, accentuated by gray bars above and below; black eyes; short, rounded ears; bushy tail with alternate brown and black rings (usually 5); body hairs long and gray

Habits Nocturnal; mainly solitary, although related females may live close to one another

Breeding Four to 6 young born around February to April after gestation period of 63 days. Weaned at 7 weeks; females usually sexually mature by their first spring, males by 2 years. May live over 17 years in captivity, up to 16 in the wild

Voice Chitters, purrs, hisses, barks, growls, snarls, and squeaks

Diet Fruit, berries, nuts, and seeds; also fish, crayfish, clams, snails, and earthworms; crops such as corn and stored grain

Habitat Almost anywhere in North America, including urban areas

Distribution Southern Canada, U.S., and Central America

Status Population: abundant. Most common member of raccoon family; continues to expand its range and increase in numbers

Common Raccoon

Procyon lotor

Raccoons are one of the most familiar North American animals. Their adaptability has allowed them to succeed in a wide range of habitats, while their appealing looks make them extremely popular with people.

THE RACCOON'S INTELLIGENCE, alertness, and curiosity were a source of fascination to early settlers and are celebrated in Native American folklore. Raccoons are often kept in captivity, and their speed of learning is thought to be somewhere between that of the rhesus monkey and the domestic cat. Young raccoons make intriguing pets, although as they mature, they can become quite a handful.

Masked Bandits

Raccoons are unmistakable animals, with their characteristic black "bandit" mask across the eyes and their bushy, banded tail. They have stout little bodies, typically weighing between 11 and 18 pounds (5 and 8 kg), although weights are known to change with season and distribution; northern animals are larger than southern ones. The heaviest raccoon recorded reached 62.4 pounds (28.3 kg).

Enormous numbers of raccoons have been trapped or shot for their skins, which are used to make jackets and hats. Raccoons are also kept in captivity to supply the fur trade. As a result of the financial gains to be made from fur farming, the common raccoon was introduced to France, the Netherlands, Germany, and parts of Russia in the 1930s and 1940s. Many escaped into the wild, and raccoons spread to Switzerland, Austria, and the Czech Republic. Some have also turned up in Poland, Hungary, Denmark, and Slovakia. The European raccoons are now sometimes considered a nuisance.

The raccoon's coat is made up of two types of hair. The short, fine underfur is uniformly gray or brownish and provides the animals with warmth and some protection from the wet. Growing from among its short coat are longer, stiffer guard hairs, which are tipped with black or white. The density of the guard hairs alters the overall appearance of the coat, often giving it a fuzzy or shaggy look. Raccoons molt in the early spring, with hair loss beginning at the head and proceeding along the back. New fur grows throughout the summer to provide extra warmth for the winter. Many raccoons have variable amounts of yellow in their coats, and some albinos have been reported. Apart from size the sexes are similar in appearance, and juveniles resemble adults.

Raccoons are excellent climbers, aided by sharp claws and the ability to rotate the hind foot through 180 degrees (thereby turning it backward). Such ability makes them one of

⊖ *The raccoon's "bandit" eye mask, brown-and-black ringed tail, and small, round ears are trademark characteristics of this highly distinctive species.*

Raccoon Currency

The raccoon's fur has always been the main reason for hunting and trapping the animal. During the 17th century bans were imposed to prevent too many raccoon skins being exported from the United States. At one time the skins were used as currency; and when the frontiersmen of Tennessee set up the state of Franklin, local officials received payments of "coonskins" each year. Although they are hard wearing, raccoon skins are not especially valuable nowadays, and trade in them is no longer a threat to population size. However, a movie about Davy Crockett, king of the wild frontier, created a sudden fashion for coonskin caps like the one worn in the movie, costing the lives of many raccoons!

only a few mammals that can hang by their hind feet and descend tree trunks head first.

Raccoons often use dens in hollow trees, preferring an entrance hole that is about 9 to 10 feet (3 m) above the ground. They also use ground burrows, brushy nests, old buildings, cellars, log piles, and haystacks in which to shelter and spend the day. In fact, they will nest almost anywhere that offers protection from predators and the weather. Each den is only used for a short period, except for over winter, when dens may be occupied for longer. When a mother has just given birth, she will also stay put, avoiding the risky and difficult business of moving her family.

Water Loving

Throughout their range raccoons are found almost everywhere that water is available. They are most abundant in forested and brushy swamps, mangroves, flood plain forests, and fresh and saltwater marshes. They are also common in cultivated and abandoned farmlands and can settle quite happily in suburban areas within parks and gardens. Raccoons are less common in dry upland woodlands, especially where pine trees grow. They also tend to avoid large open fields. Where they have spread out onto the prairies of the northern United States and southern Canada, they like to live in buildings and wet places. In desert areas they do not disperse far from rivers and springs. Raccoons are only rarely found at altitudes above 6,600 feet (2,000 m).

Raccoons do not hibernate, and in southern parts of their range they are active all year long. In the northern United States and southern Canada the coming of snow initiates periods of inactivity, although raccoons are easily roused in spells of warmer weather. During their winter sleep their heart rate does not decline, their body temperature stays above

Washing Bears

The raccoon's skillful forepaws are a prominent feature and are reflected in its name. The common name raccoon is derived from the Algonquian word *arakun*, which roughly translates as "he who scratches with his hands," a reference to the frequent grooming that raccoons characteristically indulge in. The German name *waschbär*, or "washing bear," refers to the raccoon's habit of washing its food. Even its scientific name *lotor* is taken from the Latin word *lavere*, meaning "to wash." The perception that raccoons wash themselves with their hands actually comes from observations of raccoons catching and feeding on aquatic prey. They dabble and splash in the water in an instinctive manner to catch fish, giving every appearance of washing their food. It is the same instinctive behavior displayed by captive raccoons (even when there is no water) that has encouraged the notion of the "washing bears."

The raccoon's forepaws have a well-developed sense of touch and are capable of delicate manipulation. In fact, the raccoon is almost as skillful as a monkey at handling its food.

95°F (35°C), and their metabolic rate remains high. As a result, they use more energy than true hibernators. Since they consume little or no food during their inactive period, their survival depends on the fat reserves they have built up over the previous summer and fall. In long and harsh winters raccoons may lose up to half of their body weight. Raccoons are often found denning together over winter, since they use less energy keeping warm when they snuggle up close. Up to 23 raccoons have been known to huddle together in a single den.

Raccoons are typically active from sunset to sunrise, although there is a peak in feeding activity just before midnight. Raccoons living by coastal marshes may be seen feeding during the day when their food source of crustaceans and mollusks is exposed at low tide. Raccoons are opportunists, able to make a meal from whatever food is available. It is their ability to take advantage of so many kinds of food that is the secret of their success.

In most areas plants provide the main food eaten by raccoons, especially fleshy fruit, berries, nuts, and seeds. They will also eat earthworms and insects and sometimes stored grain. Corn is a particular favorite and is usually taken just before it is ripe and ready for human consumption. Raccoons also eat small birds and sometimes snakes and lizards. Where they live near turtle nesting beaches, they will dig up and steal the buried eggs. They will also eat other vertebrates such as gophers, squirrels, shrews, rabbits, and mink; but such animals are usually already dead, so the raccoons just feed from the corpse. Raccoons occasionally scavenge the remains of larger mammals such as deer, cows, and even horses.

Breeding

Raccoons become sexually mature in their first spring, although some (particularly males) do not breed until their second season. Mating can be from December through August, occurring later in the season farther south. The peak of the breeding season is usually between February and March, with most litters being born a few weeks later in April and May. The later in the year a litter is born, the less chance the young have to fatten up for winter. Inability to survive the winter may be a factor that limits the spread of raccoons farther north.

At birth raccoons weigh about 2 to 3 ounces (60 to 75 g) and measure about 4 inches (10 cm) in length. They are covered in hair, although the mask and tail rings are represented only by dark-pigmented skin. After about three weeks they open their eyes, squirming actively and making chittering noises. Their legs become strong enough for walking by the fourth to sixth week. Their first molt occurs at seven weeks, when they shed the infant coat, and the adult fur begins to grow.

Weaning takes place from seven weeks, and the young start to leave the nest and forage for themselves. They may still be suckled by their mother for up to four months. By fall juveniles may weigh up to 15 pounds (7 kg),

⊕ Raccoons often forage beside rivers, lakes, or marshy areas where they feed on fish, crayfish, clams, snails, and other aquatic animals. They also take readily to water and are strong swimmers.

City Slickers

The raccoon's adaptability has enabled it to thrive in a variety of human-dominated environments. In fact, it has become very familiar to city dwellers. However, these masked bandits are notorious for raiding garbage bins. Not only are they known to carry away whole bins, but the nimble-fingered raiders have also learned to untie ropes used to secure the bins, rather than bite through them.

The secret of the common raccoon's success is that it can make a meal of almost any available food, including the contents of garbage baskets.

matings in his range, with the smaller males securing a few matings each.

Males may live alone or in small groups and will occupy a distinct territory ranging in size from 125 to 12,500 acres (50 to 5,000 ha). In general, raccoons will roam over 1,500 acres (600 ha) in a year. Males may travel together; but they disperse during the breeding season, when fighting and competition between them increase. Social relations are probably established and signaled through various postures, vocalizations, and scents. At least 13 different calls have been identified in raccoons. Sounds are used between individuals in close proximity to each other. Mothers keep in touch with their young by purring sounds, while hissing, short barks, and snorts express fear.

Despite the success of the common raccoon, several related species—such as the Cozumel Island raccoon (*Procyon pygmaeus*) from southeastern Mexico—are listed by the IUCN as Endangered. The Barbados raccoon (*Procyon gloverellani*) is said to have become extinct sometime after the 1960s. However, there is little to compromise the survival of the common raccoon. Predators such as wolves, bobcats, pumas, great horned owls, and alligators may pose a small threat, but few raccoons actually fall prey to them. Common raccoons are also hunted for sport (known as "coon hunting"), but relatively few are killed.

but full size is not reached until the second year. Families generally share a den over winter, and the young raccoons will leave their mother by the spring. Few wild raccoons live more than five years, but some survive up to 16 years. The oldest recorded captive raccoon was still living after 20 years.

Social Organization

The social organization of raccoons is not well known, although adults are generally solitary. However, several females—usually closely related—will live in areas that overlap, but they still tend to avoid each other. One or more males will also inhabit the same area and mate with the resident females. During the breeding season females mate with between one and four males. There is competition between males for mating privileges, with heavier males gaining greater access to the females. One successful male is likely to be responsible for over half the

The main cause for concern is the common raccoon's susceptibility to certain diseases that can be transmitted to humans, such as leptospirosis, tularemia, and most commonly (and worst of all) rabies. The common raccoon is the major carrier of rabies in the southeastern United States and in 1997 accounted for half of all reported cases of rabies from wild animals in the whole country. Raccoons also often host a type of parasitic roundworm that is harmless to the raccoon itself, but may cause death in domestic animals and even in small children.

⊙ *The common raccoon is so familiar in North America that it is often the topic of TV cartoons. A successful species, it is expanding in both range and numbers.*

⊕ *Two young raccoons by their nest in a hollow tree. As they gain independence and start to look for their own food, their mother will move them to a den at ground level to prevent them from falling.*

Common name Least weasel (European common weasel)

Scientific name *Mustela nivalis*

Family Mustelidae

Order Carnivora

Size Length head/body: 7–10 in (17–25 cm); tail length: 1–5 in (3–12 cm)

Weight 1.7–3 oz (48–85 g)

Key features Long, sleek body with short legs and short tail; flat, narrow head; fur reddish-brown in summer, with creamy-white neck and belly; turns white in winter in northern populations

Habits Solitary, territorial animals; fierce predators; very active both day and night all year round

Breeding Up to 2 litters of 1–9 young born each year after gestation period of 34–37 days. Weaned at 3–4 weeks; females sexually mature at 4 months, males at 8 months. May live up to 10 years in captivity, usually under a year in the wild

Voice Low trill to signal a friendly meeting between a male and a female; loud, harsh chirp or screech when disturbed or ready to attack

Diet Mainly small rodents, especially mice; also rabbits, lemmings, moles, pikas, birds, fish, lizards, and insects

Habitat Almost anywhere providing suitable cover and access to rodents, including meadows, farmlands, prairies, marshes, and woodlands

Distribution Northern Hemisphere: Canada, Alaska, Siberia, Japan, northern U.S., northern Europe, and Russia

Status Population: abundant. One of the more numerous small carnivores

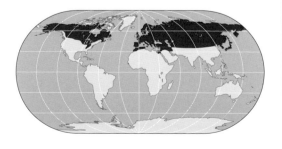

Least Weasel
Mustela nivalis

The world's smallest carnivore, the least weasel is abundant throughout a wide area of the Northern Hemisphere. However, it is an elusive creature that is hardly ever seen, remaining well hidden in dense undergrowth most of the time.

LEAST WEASELS ARE SOLITARY, elusive creatures that are hard to see, partly because they move so fast and are gone in a flash. However, they are more common than people realize. They can easily be confused with at least two other species of weasel in North America: the long-tailed weasel *(Mustela frenata)* and the stoat, or ermine *(Mustela erminea)*.

At a glance all three species look similar, but the least weasel is by far the smallest animal. The long-tailed weasel has distinct dark facial markings, and both the long-tailed weasel and the stoat have longer, bushy black-tipped tails. The European variant—the European common weasel—was once regarded as a separate species, but is now considered to be the same species as the least weasel.

World's Smallest Carnivore

Barely longer than a rat, the least weasel is the smallest carnivore in the world and the smallest of all the mustelids. It has a long, sleek body with short legs and a short tail. Its head is flat and narrow with large black eyes and prominent, rounded ears. During the summer months weasels have a reddish-brown coat with a creamy-white patch on the neck and belly. In early fall it is replaced by a lighter-colored winter coat. In some northern weasel populations, particularly in colder climates, the coat turns completely white in winter. The white color gives the animals natural camouflage against the snow and helps them avoid detection by predators.

Weasels have acute senses of sight, smell, and hearing, and often stand on their hind legs to scan their surroundings. They are incredibly

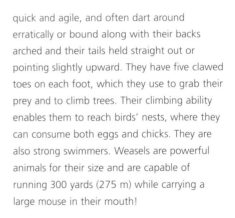

quick and agile, and often dart around erratically or bound along with their backs arched and their tails held straight out or pointing slightly upward. They have five clawed toes on each foot, which they use to grab their prey and to climb trees. Their climbing ability enables them to reach birds' nests, where they can consume both eggs and chicks. They are also strong swimmers. Weasels are powerful animals for their size and are capable of running 300 yards (275 m) while carrying a large mouse in their mouth!

Huge Appetites

Weasels are so small and dynamic and have such a fast metabolic rate that in order to survive, they must eat almost half their body weight in food each day. This means that they must catch about two mice or one fat vole per day just to stay alive. As a result, they spend a lot of time hunting, although they frequently take short rests in one of their dens.

Weasels are specialized predators of small rodents, but will also take birds, lizards, and insects whenever the opportunity arises. Their long, sleek body means that weasels are well adapted to squeezing into the smallest crevices and hunting rodents down their own burrows. In fact, the weasel's head is the widest part of its body. If it can squeeze its head into a hole, the rest of its body will follow without getting stuck. Access to such tunnels provides weasels with shelter from predators and also allows them to hunt at any time of the day or night, all year round. They do not hibernate and can hunt even under deep snow.

Weasels are renowned for being efficient killers. They catch small prey, which they kill with a few swift bites to the back of the neck. If they encounter their prey head-on in a tunnel, they kill it with a crushing bite to the windpipe. Weasels also hunt larger prey, which

⊕ *Weasels have extremely acute senses of sight, hearing, and smell. They will often stand on their hind legs to scan their surroundings.*

they stalk quietly and then pounce on the victim's back for a series of precision bites to the base of the skull. Males, which are often twice the size of females, are more likely to hunt larger prey, while the females mostly look for small rodents.

The weasel's mode of survival involves killing whatever it can, whenever it can. Faced with an abundance of mice, the voracious weasel follows the only pattern it knows and will kill more than it can eat at any one time. It sometimes stores surplus food for future meals in a side chamber off its den. Weasels are extremely versatile and can live wherever there is suitable shelter and enough food for them to reproduce successfully. They use forested, bushy, and open country, but do not normally live in wetland areas, sandy deserts, or mountainous regions. They usually make

⊕ *A weasel by a rotten log. Weasels live in a variety of habitats, including thickets and woodlands, as long as there is a good supply of suitable prey.*

their dens in rock piles, junk heaps, abandoned buildings, and burrows dug by mice, ground squirrels, or chipmunks. In colder climates they may line their nest chambers with grass or sometimes the fur and feathers of prey.

No Time to Lose

Weasels only have a short life span, but they reproduce frequently and prolifically. If food supplies are high, weasels are able to take advantage of the favorable conditions, and female weasels can have up to two litters per year. The weasels usually breed from early spring to late summer, and the pregnancy lasts about five weeks. The litter size may range from as few as one or two young to as many as 20, depending on food supplies, although an average of four to six is most common.

Newborn weasels weigh about the same as an American one cent coin and are wrinkled, pink, naked, blind, and deaf. They only open their eyes after 30 days. The mother cares

diligently for her young, which develop rapidly. By seven to eight weeks the cubs begin to accompany their mother on foraging trips and can soon kill efficiently for themselves. A few weeks later the family group begins to break up, and the young start to disperse away from their mother's home range.

Weasels are heavily dependent on rodent populations. Often in the spring, when rodent populations are low, there is an associated peak in weasel mortality, probably through starvation. However, weasels also fall victim to

predators, particularly owls and martens, but also coyotes, lynx, hawks, cats, foxes, mink, and even stoats. Weasels are also frequently killed by traffic as they dash across busy highways.

Farmer's Helpers

Least weasels are often regarded as vermin by gamekeepers and poultry farmers, and have been widely hunted and trapped. They are thought to kill young game birds but are not considered so serious a threat as stoats, which can devastate fragile populations of ground-nesting birds. In fact, weasels are often killed in traps intended for stoats.

Weasels are superbly efficient at keeping in check populations of many species of rodents that can be harmful to agriculture. One female weasel will kill hundreds of mice in a year to feed herself and her offspring. Any damage to game birds or poultry is far outweighed by the weasel's value as a destroyer of pest species, which cause untold losses to growing crops and stored food. Without predators like the weasel such losses would be even greater.

⊕ A weasel investigates a harvest mouse nest. Rodents form the bulk of the weasel's diet, and the animal is capable of crawling down burrows and squeezing into crevices in pursuit.

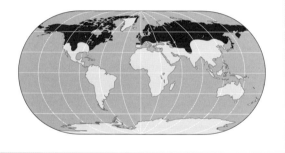

Common name
Stoat (ermine, short-tailed weasel)

Scientific name *Mustela erminea*

Family Mustelidae

Order Carnivora

Size Length head/body: 7–12.5 in (17–32 cm); tail length: 1.5–5 in (4–12 cm)

Weight 1.5–12 oz (42–340 g)

Key features Lithe, long-bodied animal with short legs and longish, black-tipped tail; body fur rich brown with cream on belly; may turn brilliant white in winter; small head with round ears, large eyes, and long whiskers

Habits Mainly nocturnal; terrestrial, but can swim and climb well; active and agile; a territorial, fierce, and solitary predator

Breeding Single litter of 3–18 (usually 4–9) young born in spring after gestation period of 10 months (including delayed implantation). Weaned at 6–8 weeks; females sexually mature at 2–3 months, males at 12 months. May live up to 10 years in captivity, many fewer in the wild

Voice Shrill squeaks when excited

Diet Carnivorous; includes small mammals, especially rodents and rabbits; also birds and eggs, reptiles and amphibians

Habitat Varied; from arctic tundra and moorland to forests and meadows

Distribution Northern Hemisphere (Eurasia and North America) from within Arctic Circle to latitude 30°N

Status Population: abundant. Common and widespread, but trapped for fur in some areas

Stoat

Mustela erminea

The stoat—or ermine, as it is called in much of its range—is the most widespread member of the mustelid family. Its range includes a wide variety of habitat from windswept arctic tundra to dense forest.

THE STOAT ALMOST CERTAINLY EVOLVED as a rodent-catching specialist, but its physical adaptations and hunting techniques make it almost equally effective in pursuit of other vertebrate prey. The animal's sinuous body is slim enough to follow rats and rabbits into burrows and to turn around inside a tunnel. Its spine is flexible, which allows it to travel much faster than its short legs might suggest. A stoat can streak through long grass at amazing speeds, take sudden leaps and bounds, and change direction in an instant. It can also climb trees and rocks and swim extremely well. Stoats have been found up to 50 feet (15 m) in trees and over half a mile (800 m) offshore in lakes. They have crossed even larger expanses of water, apparently unaided, to colonize small coastal islands in parts of their range. Stoats are light enough to run along the surface of fresh snow, but can also move below the surface, out of sight of predators and sheltered from the wind.

Regional Differences

There are as many as 29 recognized subspecies of stoat, most of them from North America, where they are often known as short-tailed weasels. They are distinguished as much by geography as by any obvious physical characteristics. As a general rule, American stoats are smaller than those in the Old World, and throughout the stoat's geographical range males are bigger than females, sometimes twice the size. Perhaps the most notable regional difference is that in higher latitudes stoats turn white in winter, while those in more temperate zones retain their brown color all year round.

Stoats are small, but they sometimes use surprisingly large home ranges. An active male

⬇ *Stoats carry out all their explorations at high speed and not surprisingly burn up a great deal of energy. The long, thin body is also inefficient when it comes to conserving heat.*

will usually occupy a range of between 50 and 150 acres (20 and 60 ha), although ranges up to 500 acres (200 ha) have been recorded. Male ranges overlap partially with those of females, which are usually about half the size of the area used by males. Within their ranges both males and females maintain an area of private territory, inside which other stoats are not tolerated. Territories are marked out with scent and droppings placed on landmarks such as rocks and tree stumps. The resident stoat will regularly patrol its home area. Hunting and patrolling normally take place between dusk and dawn, and a stoat may cover anything up to 10 miles (16 km) a night. Both males and females use dens for sleeping during the day—there are usually several within the home range, located in rock crevices or rodent burrows.

Frenetic Activity

A stoat's heart beats up to 500 times a minute, and almost all activities are carried out at a similarly frenetic pace. Stoats are incredibly inquisitive and will investigate any nook or cranny within their range, darting in and out of burrows, tree holes, and crevices, nose twitching and ears pricked for the scent or sound of potential prey.

A stoat emerges from the shelter of a log. Scent and droppings are deposited on logs and rocks by both males and females to mark out a home range.

Stoats have excellent eyesight, but in the darkness underground they rely heavily on their long, sensitive whiskers to find their way around. Stoats do not hibernate, so they have to remain active and well fed right through the winter. Where possible, a stoat will make provision for the cold season by hiding caches (stores) of spare meat caught earlier in the year. The fact that stoats will kill more than they can eat means they are sometimes accused of killing for fun, when in fact they may be just thinking ahead. Given the chance, all the excess will be carefully stockpiled for later.

Polygamous Relationships

Male stoats are larger than females, and as such are sometimes at a disadvantage when hunting because they cannot pursue small prey such as mice and voles into such tight spaces. However, large size comes into its own when defending a territory, within which the resident male will have by far the best chances of mating with all the local females. Once mating is over, the male moves on almost immediately in search of another female and plays no further part in the rearing of his future family. In fact, stoat pregnancy is so long that more often than not the father will be dead by the time his kits are born, some 10 months after mating. Like many other mustelids, stoat gestation includes a nine-month period of delayed implantation, during which embryonic development is halted at an early stage. Development resumes the following spring, and

Winter White

All stoats undergo two molts each year, in the spring and fall. The timing of the molt is determined by the changing day length. For example, in the fall the change to a winter coat is triggered when the daylight hours drop below a critical limit. Hence the molt happens sooner in the north, where the winter days are shorter. While molting is controlled by day length, the color of the winter coat is determined by air temperature. However short the days, if the average temperature remains above a certain level, the new winter coat will be brown. Below that temperature it will be white, providing the stoat with useful camouflage in a climate where snow is likely. In areas where the temperature fluctuates around the threshold level, the coat will be a mixture of brown and white. The molt from a summer to winter coat begins on the belly and moves forward to the head. In spring the process happens in reverse.

In the most northerly parts of its range the stoat's winter pelt is highly prized for its fineness and purity of color. Despite being trapped for their fur, stoats are among the world's most successful predatory species.

↩ *A stoat uses its keen sense of hearing and smell to detect prey. Its sinuous body allows it to pursue rats and rabbits into their burrows and to turn around in tight spaces with ease.*

litters of up to 13 babies are born. They are tiny, blind, and virtually naked, apart from a fine covering of wispy white hairs. The family is raised in a den lined with the fur and feathers of prey animals, and defended fiercely by the mother. If she is disturbed, she will often move the family to a new, safe place, carrying them one at a time in her mouth.

The young develop quickly. By six weeks they are as heavy as their mother and begin to accompany her on hunting excursions. She teaches them the necessary techniques and

gives them dead or maimed prey animals to practice on. Stoats are unusual in that they will kill animals larger than themselves without help from other members of their species.

Stoats are among several European animals to have been introduced to New Zealand by settlers. By the early 19th century New Zealand already had a serious problem with introduced mammals, especially rabbits. Such animals had been taken there some years before to provide meat, fur, and sport hunting. As in Australia, the New Zealand rabbit population rapidly grew out of control, and the introduction of stoats was a misguided attempt to curb their numbers. Although stoats did kill rabbits, the whole operation was a disaster. The stoats soon found that it was much easier to kill New Zealand's native birds, which—in the absence of native mammalian predators—had no instinctive fear of the stoat and in many cases were flightless. By the turn of the century stoats had overrun both North and South Islands. They remain widespread even today despite rigorous attempts to control them.

Common name American marten (American pine marten, American sable)

Scientific name *Martes americana*

Family Mustelidae

Order Carnivora

Size Length head/body: 20–27 in (50–68 cm); tail length: 7–9 in (18–23 cm)

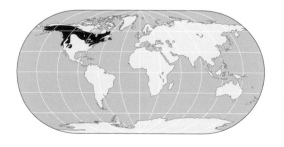

Weight 10–44 oz (280–1,250 g). Male generally at least a third bigger than female

Key features Slender cat-sized animal with short legs and bushy tail; fur ranges from pale brown to almost black with an orange or yellowish throat patch

Habits Active throughout the year and also at any time of day or night; climbs well

Breeding Up to 5 (usually 3) young born once a year in the spring after gestation period of 1 month. Weaned at 6 weeks; sexually mature at 15–24 months. May live 17 years in captivity, at least 15 in the wild

Voice Normally silent; sometimes makes chuckling noises or gives the occasional scream

Diet Wide variety of small animals; also insects, fruit, and seeds

Habitat Deciduous and coniferous forests

Distribution From Alaska eastward to eastern Canada; also Rocky Mountains and Sierra Nevada

Status Population: abundant. Widespread but elusive; rare and declining in some places

American Marten

Martes americana

A charismatic small forest predator, the American marten has suffered heavily from the activities of trappers and also from habitat loss.

AMERICAN MARTENS ARE TYPICALLY associated with northern forests, particularly those dominated by mature spruce and fir trees. They also occur in deciduous forests, but the most distinctive features of their habitat are the complex mixture of different species, the different ages of the trees, and the presence of glades and clearings. Such habitat generally has abundant and diverse prey. The martens will move around using different areas at different times of year, depending on food availability.

Disappearing Habitats

Martens are found across much of North America from Alaska through the forested parts of Canada and east into the eastern United States. At one time they were also fairly common in the southeastern United States. However, the harvesting of trees and clearing of forests to make way for farmland have since deprived the animals of enormous areas of suitable habitat. Moreover, the dense and lustrous fur of the marten made it a prime target for fur trappers, and the animals were easy to catch. As a result, martens were eliminated from the southern part of their natural range and have become quite rare in many other parts of their range, too.

American martens do not hibernate and are active throughout the year. They hunt mainly on the ground, although they can also climb well. They are perfectly capable of swimming, but do not enter the water unless absolutely necessary, preferring to cross streams by way of overhanging trees and logs.

American martens have a very varied diet, and more than 100 different types of food have substances—precisely to deter animals such as martens. However, the American marten is not wholly carnivorous and will also eat large quantities of fruit in the fall and even ripe seeds. The indigestible remains of its varied diet can be seen as chewed fragments in its shiny black feces, which are often deposited on open tracks and prominent logs, probably to help mark out its territory.

Martens are extremely active creatures and can move rapidly when they need to. They trot and bound around, stopping frequently to investigate likely places to find food. Most of their foraging is done early in the day and late in the evening, although martens may be active at any time of the day or night.

Delayed Implantation

Martens are territorial, and each one lives alone, except during the breeding season. A male may live with a chosen female for a couple of weeks in the summer, during which time the animals will play together and indulge in mock fights. They mate many times, often with several partners in a season. The pregnancy lasts only a month, but births in midwinter are avoided by a process known as delayed implantation. The fertilized egg remains dormant for over 200 days before implanting itself in the wall of the uterus and developing normally. As a result, the young are not born until the following April, when there is plenty of food, and the weather is less challenging.

Litters average three young, but sometimes there are as many as five babies in the family. The offspring are born in a den among boulders, a hollow tree, or the shelter of a fallen log. They grow rapidly, and by midsummer they are almost as large as their mother. Males will grow to at least one-third bigger than females. The young face few predators and are nimble enough to escape likely attackers. However, winter is a difficult time when it is hard to find mammal prey under deep snow. Nevertheless, survival rates are high, and some martens live for up to 15 years.

⬅ The American marten is becoming quite rare in places as a result of the loss of its forest habitat to farming and logging.

been reported. Their main prey consists of small mammals such as red-backed voles. However, they will also eat mice and occasionally larger animals, such as chipmunks, ground squirrels, pocket gophers, and even snowshoe hares. Their preference for voles and mice is due to the fact that such prey is common and easy to catch in the long grass of forest clearings. Martens will also eat birds if they can catch them and any large-bodied insects, except those that produce evil-tasting

Common name American mink

Scientific name *Mustela vison*

Family	Mustelidae
Order	Carnivora
Size	Length head/body: 12–18.5 in (30–47 cm); tail length: 5–9 in (13–23 cm)
Weight	1.9–4 lb (0.9–1.8 kg); female 1–1.8 lb (0.5–0.8 kg)

Key features Resembles short-legged, glossy black or dark-brown cat; pointed muzzle

Habits Mainly nocturnal; swims and dives; uses burrows and lairs among tree roots at water's edge, also rabbit burrows, but does not dig for itself

Breeding One litter of 4–6 young born April–May after gestation period of 39–78 days, including a variable period of delayed implantation. Weaned at 5–6 weeks; sexually mature at 2 years. May live for 10 years in captivity, 2–3 in the wild

Voice Hisses when threatened; may scream defiantly in self-defense, but usually silent

Diet Fish, frogs, small mammals, waterside birds and their eggs; also some invertebrates such as beetles and worms, especially along coasts

Habitat Mostly lowland areas beside rivers, lakes, and ponds; also marshland and along seashores

Distribution Canada; eastern and most of central U.S.; introduced to Europe: in Britain, France, Italy, Spain, Ireland, Scandinavia, and Iceland

Status Population: abundant. Increasing in Europe

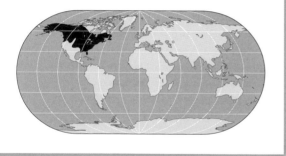

American Mink

Mustela vison

A widespread waterside predator in North America, the American mink has also become established in parts of Europe, where it is proving a successful but unwelcome invader.

AMERICAN MINK ARE WIDESPREAD and fairly common across most of North America. They are substantially smaller than otters and much darker in color, appearing almost black, especially when wet. Their droppings are easily recognized, being black and cylindrical and about the diameter of a pencil. They are deposited on rocks and logs at the water's edge and smell distinctly unpleasant.

Mink are usually associated with slow-flowing rivers and lowland lakes, preferring places where there is plenty of overhanging bankside vegetation. Each has a hunting territory of about half a mile of riverbank or lakeshore. Mink are less common in upland areas, but in some places they have established colonies along the coast. Here they can reach high densities, as many as three individuals per mile (1.6 km) of coastline. Along the seashore mink behave rather like coastal otters, feeding on rockpool fish, but also climbing steep grassy cliff slopes to raid gulls' nests.

Lone Rangers

Mink are active at dusk and after dark, bounding around on land and swimming and diving underwater. They are unsociable creatures, and each one tends to live in its own territory. Male territories do not overlap, but often include parts of the territories of one or more females. In spring some of the males set out on long journeys in search of females and may travel widely across the countryside, helping colonize new areas. However, even after finding a mate mink still do not set up families. There is no pair bond, and after mating the animals live apart. Females produce only one litter each year, usually in April or May. The average litter size is four to six young, although captive mink can produce many more. The babies spend up to two months in their mother's nest, normally in a burrow or among dense tree roots. They are weaned at about five to six weeks, and the family disperses soon afterward. The babies grow rapidly to reach adult size before the end of the year and are capable of breeding the following spring. Females can still breed at seven years, although few reach this age in the wild, and most mink die within the first three years of life.

Dangers Faced

Young mink tend to disperse away from their mother's territory, sometimes traveling more than 6 miles (10 km) in their search for a place to live. Once they have established a new territory, mink tend to stay in the same place, often for several years. They need to know their home patch well, since they are in constant danger from gamekeepers, farmers, and trappers. Thousands are killed each year. Others may drown in traps set to catch fish. On the other hand, mink have little to fear from natural predators and are well able to defend themselves if attacked by foxes or cats. Perhaps that is why they are often active in broad daylight, although their main activity takes place after dark, when prey animals are more likely to be found.

Mink eat fish and birds, but also take beetles, worms, and other invertebrates. Most of their prey consists of small creatures, but often mink will attack rabbits. In fact, they will eat almost anything apart from fruit and other plant material. Mink are such successful hunters they spend less than 20 percent of their time away from the den. The rest of the time they are safely tucked away asleep or grooming their

⊙ *Mink are solitary creatures and only meet to breed. However, after breeding, they do not set up families, and the female raises the young alone.*

produced, ranging from silver-white to cream, as well as the natural dark chocolate-brown.

The mink industry expanded in Britain and Europe during the 1950s, producing more than a quarter of a million skins per year. But the mink is able to climb and squeeze through small gaps, and inevitably many escaped. The animals often proved adaptable and survived successfully in the wild, despite attempts to eradicate them. Mink spread widely and within 30 years occupied most of Britain. They were also found over most of Iceland, Norway, and Sweden. Today they continue to spread eastward in mainland Europe and are now also established in Ireland.

Invaders of Europe

The American mink's extraordinarily successful invasion has added another species to the mammals of Europe, but its spread has been accompanied by serious losses of native animals. The rare European mink (*Mustela lutreola*) has now become almost extinct, and in Britain the water vole has disappeared from nearly 90 percent of sites where it was formerly common. Mink are good swimmers and have managed to reach islands over 2 miles (3 km) offshore where seabirds had previously nested in safety, protected by the surrounding water. Here they have killed hundreds of adult birds and chicks, as well as eating the eggs. Whole colonies have been wiped out in the space of only a few breeding seasons, and in some places seabirds have been almost eliminated.

When mink get into parks with captive ducks (often pinioned so they cannot fly), they cause mayhem, killing many birds. They also create havoc on chicken farms and fish farms where there are high densities of juicy trout to eat. Gamekeepers regard mink as vermin because they take many gamebird chicks and eggs. However, recent studies suggest that mink numbers may be falling in Britain as otters recolonize rivers, perhaps displacing the smaller mink. In the United States mink seem less of a problem, so perhaps one day they will achieve a better balance with nature in Europe, too.

sleek and lustrous fur. The den may be in an old rabbit burrow among rocks or piles of brushwood, but it is always close to the water's edge. It may sometimes have a separate entrance underwater. Each mink may use several dens at different times of the year. Mink do not hibernate, although they become significantly less active during winter months.

Fur Farming

Mink have a glossy coat that has been highly prized by the fashion industry. Although wild mink are easy to catch, trappers could not obtain enough skins to meet demand. As a result, special mink farms were established, especially in Europe. Mink were first imported from America to European fur farms in the 1920s, but only small numbers were kept until after the Second World War. However, in the postwar years raising mink for the fur trade was seen as a lucrative new moneymaker. The animals bred well in captivity and could be fed cheaply on unwanted animal waste, including bits of chicken from the expanding broiler fowl industry. Moreover, through careful selective breeding even more valuable colored furs were

A fur farm in Estonia. Breeding mink for the fur trade was once a lucrative business, and a mink coat was considered the height of luxury. However, fur has since fallen out of fashion.

The Price of Freedom

In 1998 animal rights supporters broke into British fur farms and released thousands of mink. Whether fur farming is right or wrong, it was an irresponsible act, widely condemned by conservationists and animal welfare groups. The released mink posed a serious threat to many other species, and such releases were also cruel to the mink themselves, since they had been bred in captivity for many generations: Like pet mice or guinea pigs, they were unaccustomed to life in the wild. Large numbers were easily recaptured, since they had no idea how to avoid being caught. Many were also run over on the roads. Others were killed by dogs or shot by annoyed landowners and gamekeepers. Some mink even found their way into people's houses, having been driven by hunger to enter through the cat flap in search of food.

Fashions have changed, and fur coats are no longer so much in demand. Mink farming is less profitable today than it once was, and animal welfare legislation imposes many conditions on the managers of fur farms. Large numbers of businesses have closed down, leaving fewer mink in captivity. However, populations of wild American mink in Europe continue to prosper and are likely to remain permanently established there.

⊖ *An American mink returns to a henhouse in Britain to feed on hens it killed the previous night. Mink have also been responsible for losses to many native animals.*

European Mink

The native European mink *(Mustela lutreola)* is similar in appearance to its American cousin, but has a white area around its upper lip. European mink prefer to live beside fast-flowing water rather than lakes and coast, but have become notably scarce in recent times. The species is now found only in small areas of France, Spain, Estonia, and Romania, having been driven out of the rest of Europe by habitat loss and the invading American mink. It has disappeared from about 80 percent of its former range and is still in decline. Urgent efforts are being made to establish captive-breeding colonies from which the species might one day be restored to the wild.

The rare European mink (above) has a white patch around its upper lip.

Common name Wolverine (glutton, skunk bear)

Scientific name *Gulo gulo*

Family	Mustelidae
Order	Carnivora
Size	Length head/body: 24–26 in (62–67 cm); tail length: 5–10 in (13–25 cm); height at shoulder: 14–17 in (35–43 cm)
Weight	20–65 lb (9–29 kg). Male at least 10% bigger than female
Key features	Low, thickset animal with short legs and large, powerful paws; coat dark brown but paler on face and flanks; tail thick and bushy
Habits	Solitary creature that roams widely; mainly nocturnal in summer
Breeding	One litter of up to 4 babies born February–March after gestation period of 30–40 days (including up to 9 months delayed implantation). Weaned at 8–10 weeks; sexually mature at 2–3 years. May live up to 18 years in captivity, 11 in the wild
Voice	Hisses and growls when annoyed; also playful squeaks and grunts
Diet	Mostly rodents; sometimes larger mammals, especially as carrion; also fruit, berries, birds, and eggs
Habitat	Mountainous forests, rocky areas, and tundra in summer
Distribution	Widely distributed across northern Europe and Russia; also Canada and northern U.S.
Status	Population: unknown, probably low thousands; IUCN Vulnerable

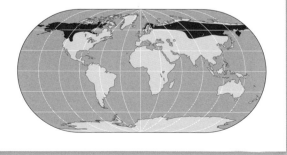

Wolverine

Gulo gulo

A naturally scarce animal, the wolverine has a reputation as a fierce and greedy predator. Ruthless persecution has resulted in it becoming quite rare.

WOLVERINES ARE FOUND AROUND the edges of the Arctic, in both North America and Europe. The cold conditions and long winters mean that plants grow slowly, and prey animals are scarce. As a result, animals that live in the Arctic all year—particularly predators like the wolverine—tend to be low in number. Wolverines need huge areas to provide them with enough food. They must wander widely to feed, sometimes covering over 30 miles (48 km) in a day. Often there is only one wolverine in more than 200 square miles (500 sq. km).

The summer months may be spent out on open tundra or mountainsides, but come the winter wolverines will migrate to the shelter of forests, sometimes journeying up to 50 miles (80 km). Although they can cope with deep snow, if there is too much of it, the short-legged wolverines have difficulty chasing and capturing prey. As a result, winter is normally spent in the relative shelter of conifer forests, where the snow is usually less deep.

Waste Not, Want Not

Although wolverines normally prey on small creatures, such as Arctic hares and lemmings, they must make the most of any food they manage to find. In the fall there are plenty of ripe berries to eat and many inexperienced young mammals and birds that are easy to kill. During leaner times the wolverine will sometimes attack animals as large as deer and wild sheep, storing surplus food for later. (The wolverine's powerful jaws mean it can drag down animals much larger than itself.) The wolverine's tendency to kill more than it needs has given rise to its reputation as a wanton killer. There are many folktales about the cunning and bloodthirstiness of wolverines. Fur

⊕ *Not unlike a small bear in appearance, the wolverine is the largest member of the weasel family. It is territorial and will defend its home patch against others of the same sex.*

trappers complain that the animals steal from their traps, and reindeer herdsmen fear that wolverines will attack their animals.

Wolverines mate in the summer and would normally give birth after a few weeks. However, they avoid producing young in midwinter by a process known as delayed implantation. As a result, the babies are not born until about April, just in time for the long days and relatively abundant food supplies of the Arctic summer. The young continue to use the mother's den until May. They stay within her normal home range until about August. After that they tend to disperse, especially the males.

The Human Threat

Many wolverines are shot on sight, not just because of the damage they might cause, but also for their fur. Their long hairs tend not to freeze together in the intense cold of the Arctic, so the pelts are prized by local people for trimming the hoods of winter coats.

Wolverines are sensitive creatures and easily disturbed. The remote areas in which they once lived in safety are becoming more accessible. The remaining populations of wolverines are regarded as threatened, and the species seems likely to suffer further decline. Wolverines were at one time found much farther south in Europe than they are today, even into Germany. In the United States they used to occur across the whole continent as far south as Arizona and New Mexico. Today they are extinct east of Montana, with a few remaining in California and Idaho.

Wolverines are still widespread in Canada, and British Columbia is a stronghold with about 5,000 animals. Little is known about numbers in Russia and Siberia, but in northern Scandinavia wolverines have become rare and now occur mainly in the remote mountains of Norway and Sweden. It is thought that only 40 are left in the whole of Finland.

⊕ *The wolverine's broad feet allow it to run on snow without sinking in.*

North American River Otter

Lontra canadensis

North American river otters are playful, intelligent animals that were once common throughout most of the United States and Canada. Many rivers are now too polluted or urbanized to support them.

Common name North American river otter (northern river otter)

Scientific name *Lontra canadensis*

Family Mustelidae

Order Carnivora

Size Length head/body: 26–42 in (66–107 cm); tail length: 12–20 in (32–46 cm); height at shoulder: 10–12 in (25–30 cm). Male larger than female

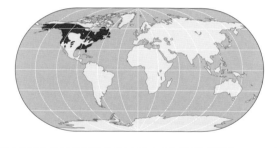

Weight 11–31 lb (5–14 kg)

Key features Long, cylindrical body and short, stocky legs; long, pointed tail; small, blunt head with small ears and eyes; fur light to dark brown

Habits Lively and playful: lives alone or in small groups; semiaquatic; active at any time of day

Breeding Litters of 1–6 (usually 2 or 3) cubs born from November–May after gestation period of approximately 50 days (including delayed implantation). Weaned at 5–6 months; sexually mature at 2 or 3 years. May live to 21 years in captivity, 14 in the wild

Voice Shrill chirps, soft "chuckles," grunts, coughs, and growls; loud screams when frightened

Diet Mostly fish; also crayfish, frogs, crabs, birds' eggs, and small mammals

Habitat Coastal and freshwater: rivers, streams, lakes, reservoirs, marshes, swamps, and estuaries

Distribution Canada and mainly northwestern, southeastern, and Great Lakes states of U.S.

Status Population: probably low thousands; CITES II. Common in some areas of U.S. and Canada, but extinct or rare in others

NORTH AMERICAN RIVER OTTERS used to be found in all the major waterways of the United States and Canada until at least the 18th century. Along with the beaver and timber wolf, otters were the most widely distributed wild mammals in the United States. They favor coastal areas, estuaries, rivers, lakes, and streams—in fact, any healthy water systems that can provide them with plenty of fish. Yet they are shy animals and do not like disturbance. When waterways are developed for housing, industry, or recreation, the otters usually leave. Even away from cities and towns agriculture poses a threat to otter habitats by removing tree and scrub cover and polluting water with pesticides. Because otters are at the top of the food chain, they are sensitive to any pollutants. Another threat to otter survival is the fur trade. About 20,000 to 30,000 animals are harvested every year. Such large numbers may actually be sustainable as long as the otters' habitat is healthy. Sadly, this is frequently not the case, and pollution, disturbance, and loss of habitat have brought about the extinction of otters in many areas.

Favorite Games

Otters are intelligent, quick, curious, and exuberant creatures. They are famous for their playful nature and will make a game of any activity. The young and even adults will play games of tag, tumbling, and wrestling. A favorite pastime is sliding down mud or snow banks into the water, tobogganing on their belly with their front legs folded back. Small objects such as shells, stones, or sticks are used as toys and for games of hide and seek. In captivity otters have been shown to have good memories and can be trained to perform tricks such as retrieving objects.

American river otters are well adapted to their aquatic life. They have a long, streamlined body and feet with webbed toes. The fur is dense and oily, with long, glossy guard hairs. Long, sensitive whiskers help the otter locate prey in murky water, and the small eyes are set high on the head and close to the nose, so the otter can look around above water while keeping most of its body submerged.

Otters are very graceful in the water. They swim by twisting their hind quarters and tail from side to side. They have a top speed of 6 to 7 miles per hour (10 to 11

⊕ North American river otters are not as widespread as they once were and are now mostly confined to Canada and the northwestern, southeastern, and Great Lakes states of the United States.

km/h) and can dive to at least 60 feet (18 m) deep. They can stay submerged for over four minutes, but most dives are shallow and brief.

Waterway Travelers

Otters can have very large home ranges and may use many miles of waterway. Males generally travel farther than females. The length of a home range can vary from 5 to 50 miles (8 to 80 km), with the size depending on habitat quality. An otter living in an area with plenty of food, for example, will not have to travel as far as one that lives where prey is scarce. Territories of individual animals may overlap, and the home range of a male usually crosses that of one or more females.

Within their territories American river otters have many resting sites and dens. Rather than digging their own dens, they usually use holes

made by other animals, such as beavers. They otherwise use natural shelters such as hollow tree trunks, driftwood piles, or even abandoned boathouses. Nursing dens have an underwater entrance with a tunnel leading to a nesting chamber lined with leaves, grasses, mosses, and hair to provide a soft bed for the young.

Scent Marking

Otters mark their territory using scent. Like other members of the weasel family, an otter has scent glands near its anus that produce a strong-smelling musk, which the animal wipes onto scent posts throughout its territory. They may be tree stumps, prominent stones, or logs, usually well above the water line so that scent marks do not get washed off by rising water levels after rain. As well as depositing musk, otters leave spraints (feces), which provide messages for the inquisitive noses of otters who visit later. Scent posts are usually in obvious places on the otter's route, including dens, rolling places, slides, runways, and haul-outs. Otters will also scratch up mounds of soil and debris or twist tufts of grass together, marking them with scent deposits or spraint. As well as defining boundaries, scent marking also signals when the otter last passed through, its sex, and probably its age, helping avoid potential conflict with other otters.

Although generally solitary, otters will spend time in a small group, which can be made up of a mother and her pups, a male and

⊕ A river otter with young by a stream in Montana. The young are born blind and helpless, but fully furred. They are introduced to water at about seven weeks, but are often reluctant swimmers and have to be dragged there.

female together, or even a group of bachelor males. Groups are temporary, have no apparent leader, and do not cooperate in hunting or share what is caught.

Most American river otters start to breed when they are two years old. Receptive females advertise their condition by markings at scent stations, and the powerful smell may attract two or three males. Although American river otters do not form strong pair bonds, once a male has mated, he may drive away other males who come near the female while she is receptive. He will then leave her and takes no part in rearing the offspring. However, sometimes fathers have been known to rejoin the family group when the young are about six months old. The young are born between November and May, but usually in March or April. The pups are born blind and helpless, but fully furred, and look like miniature versions of the adults.

Reluctant Swimmers

Otter milk is very rich, and the young grow quickly on it. They open their eyes at about four weeks and begin playing with each other and their mother. The mother introduces the pups to water by the time they are seven weeks old: They are often reluctant swimmers and may need to be dragged into the water. A mother

will spend a lot of time teaching her pups, catching small fish and releasing them again so the young otters can develop their hunting skills. By nine or 10 weeks of age they start to eat solid food, although they will not be fully weaned until they are at least three months old. They are not fully independent for a further six months, but sometimes members of a family will stay together for a year or two.

Otters eat lots of fish, along with smaller amounts of other prey. They consume up to 2 pounds (1 kg) of food in a single meal. They have a rapid metabolism, so they need to eat frequently. When hunting, they spend a lot of time diving and chasing fish or digging in mud and stones at the bottom of ponds and streams for smaller prey. Crayfish are an important part of the diet, as are crabs along the seashore. Otters catch prey in the mouth, not the paws. Small food items are eaten in the water, while larger ones are taken ashore.

Otters tend to catch large or slow-moving fish. During salmon spawning times otters feast on the hordes of fish concentrated into small streams or shallow pools with nowhere to escape—exhausted "spawned-out" salmon are easy to catch. For most of the year, however, otters depend on other types of fish. Although fast-swimming species such as trout and pike are common in rivers and lakes, otters usually

Signs of Otter Activity

Otters are shy and elusive; few people are lucky enough to see them. However, it is relatively easy to find clues to their activity. When they come out of the water, they roll on the bank to dry themselves, leaving large areas of flattened vegetation. In soft mud their footprints show the distinctive marks of the webs between their toes. Spraints may also be visible—sometimes on twisted tufts of grass or piles of dirt and vegetation.

go for easier prey. They do not deserve the bad name given by anglers: Otters eat some sport fish, but are more likely to eat other species that actually compete with sport fish for food.

Otters will sometimes eat waterfowl, such as coots and ducks, and raid their nests for eggs. They may take dead or injured birds, particularly in the shooting season, but will also actively hunt and kill healthy birds. They stalk them by swimming underwater and grabbing the birds from below. Otters are also among the few predators that kill snapping turtles. Occasionally, otters will stalk birds and mammals on land. There have been reports of American river otters chasing and catching small mammals up to the size of snowshoe hares. They will also eat berries, such as rosehips and blueberries, but that is unusual.

⊕ *An otter feeds on a trout caught in a river. Generally, river otters will take large or slow-moving fish rather than fast-swimming species, such as trout. During salmon spawning times otters will feast on exhausted fish resting in shallow pools.*

Common name Sea otter

Scientific name *Enhydra lutris*

Family	Mustelidae
Order	Carnivora
Size	Length head/body: 29.5–35 in (75–90 cm); tail length: 11–12.5 in (28–32 cm); height at shoulder: 8–10 in (20–25 cm)
	Weight 30–85 lb (14–38 kg)
Key features	Dark-brown otter with blunt-looking head that turns pale cream with age; feet completely webbed; hind feet form flippers
Habits	Floats on back in kelp beds and calm waters; dives to feed from seabed
Breeding	One pup born each year in early summer after gestation period of 4 months (including up to 8 months delayed implantation). Weaned at 5 months; females sexually mature at 3 years; males at 5-6 years, but do not breed successfully until at least 7 years. May live for over 20 years in captivity, similar in the wild
Voice	Normally silent
Diet	Crabs, shellfish, sea urchins, fish, and other marine animals
Habitat	Kelp beds and rocky seashores
Distribution	Formerly along coasts across eastern and northern Pacific from California to Kamchatka and northern Japan; exterminated over most of its range, now reintroduced to coasts of California, Alaska, Oregon, and Washington
Status	Population: about 150,000 and growing; IUCN Endangered; CITES II. Given full legal protection in 1911 and probably now secure

reintroduced population
in North America

Sea Otter

Enhydra lutris

The sea otter was once widespread along the coasts of the North Pacific, but hunting for skins brought the species to the brink of extinction. It has now substantially recovered, thanks to strict international protection and some successful reintroductions.

TODAY SEA OTTERS ARE EASILY observed, especially along the California coast, and their playful antics make them a popular species to watch. They spend most of their time floating quietly on their back at the water's surface, grooming their fur and rolling over and over in the waves. They also spend long periods dozing on their backs—often anchored by a strand of kelp draped across their chest. Periodically, they feed by making short dives of about a minute to the seabed to look for crabs, sea urchins, and mollusks. They cannot dive deeply, so they have to stay in relatively shallow waters. They also cannot last long without food and are unable to make long journeys out to sea if it entails crossing large areas of deep water. However, they do sometimes undertake long journeys along the coast, sticking close to shore where they can continue to feed along the way. Generally, sea otters are solitary animals, and males are territorial, probably to avoid competing for limited food resources.

Unwitting Conservationists

Sea otters are intelligent animals and have learned to bite open old drink cans that have sunk to the bottom of the ocean and now provide a lair in which a small octopus may hide. They also eat large numbers of sea urchins, helping control their numbers. Keeping numbers down is important because the urchins eat growing kelp. If there are too many urchins, the kelp beds are unable to flourish and do not protect the coast from the full force of the Pacific tides. Beach erosion and flooding may result. Hence sea otters are very important ecologically for maintaining a healthy coastline.

⊕ *The sea otter possesses the densest coat of any mammal to help keep it warm in the chilly seas of the North Pacific. There are estimated to be over half a million hairs per square inch of fur.*

Densest Fur on Earth

Sea otters are probably the smallest warm-blooded animals that spend all their time in the water. The coastal seas of the North Pacific are very cold: Even far south off California the sea is cool and will chill a mammal's body quite quickly. The sea otter therefore needs very effective insulation to prevent loss of body heat. Its protection is provided by a thick coat of the densest fur possessed by any mammal. There are estimated to be more than half a million hairs per square inch on the sea otter's body—twice as many as found on the larger fur seals. The fur has long, shiny guard hairs that help keep the water at bay and prevent the underfur from becoming matted and losing its insulation value. Below the guard hairs is a dense mass of extremely fine underfur that traps a layer of air against the skin and acts as insulation to prevent heat loss.

The sea otter is totally dependent on its fabulous fur to enable it to live in the cold seas without becoming chilled. It therefore spends much of its time grooming and caring for its precious coat. That is also why the animal cannot dive deeply, since the increased water

ⓐ Sea otters make dives to the seabed to look for prey such as crabs, sea urchins, and mollusks. However, they easily get out of breath, so cannot dive deeply, and must stay in shallower waters along the coast.

pressure at depth squeezes the vital air from the fur, causing it to lose its insulating properties. The trapped air also makes the otter rather buoyant, so it has to expend more energy swimming down into the water than would be needed by a small seal or whale. It soon gets out of breath, which prevents it from staying underwater long enough to reach greater depths. The otter's fur is not just valuable to the animal: For over 200 years it was also one of the world's most prized furs for human use, making warm coats for winter wear. A sea otter's skin could be sold for the equivalent of an entire year's wages for a seaman, so there was plenty of incentive to hunt the otters. Hunting sea otters and fur seals was one of the main reasons for the early exploration of the North Pacific. During the 18th century Russian navigators expanded their trade in skins and colonized Alaska and what is now British Columbia, as well as the Aleutian Islands. Later the British and Americans joined in.

Skin Trade

Over three-quarters of a million sea otters were killed between 1750 and 1850, with 17,000 skins in a single shipment made in 1803. Pelts were bartered with the Chinese in exchange for fine porcelain, which could then be taken to Europe and sold for immense profits. Expeditions would stock up with axes and other useful tools made in Europe and North America, then sail back to the North Pacific and trade the tools with local hunters for yet more sea otter pelts.

The otters were easily hunted from canoes and kayaks. An animal would be chased so it was forced to dive repeatedly until it was out of breath. When it was too exhausted to dive any more, the hunter speared it and dragged the body into the kayak, where it was skinned. The body was then thrown back into the sea, and the hunter paddled on to find the next otter. Since the otters only lived along the coast and did not seek safety by dispersing out at sea, they could be hunted systematically until every last one had been killed along hundreds of miles of coastline. Sea otters have few natural predators (killer whales, bears, and bald eagles occasionally kill them) and are not adapted to withstand heavy losses. They do not breed rapidly and produce only one youngster per year—often not even every year. Females do not breed until they are over three years old and sometimes can be over five years old before producing their first baby. Male sea otters take even longer to mature and may not have a breeding territory established until they are 10 years old.

Slow breeding meant that sea otter populations could not cope with heavy exploitation, and the animals soon disappeared over wide areas. By the early 20th century the sea otter had become exceedingly rare, having been reduced to perhaps fewer than a thousand animals in the whole North Pacific. It was on the brink of total extinction. Yet its range crossed several international borders so giving it legal protection in one country would not necessarily help: The animals might be killed illegally in one country, but then smuggled out to be sold somewhere else. What was needed was an international agreement to give the animal legal protection everywhere. However, that had never been done before for any marine animal. In 1911, in the first such international agreement, the Russians, Americans, and British (on behalf of Canada) agreed to total protection for the sea otter throughout the North Pacific.

⊕ *A sea otter feeds on shellfish while swimming on its back. Sea otters carry a flat stone to help them smash open hard shells. Prey is placed on the otter's stomach and crushed with the stone.*

Tool User

The sea otter is one of the few animals, apart from apes, that has learned to use tools. It often carries a flat stone tucked into its armpit and uses it to help smash open the hard shells of the crabs, mollusks, and sea urchins on which it feeds. The otter lies on its back, floating in calm water, with its prey lying on its chest. The animal uses its paws to lift the stone and hit the prey hard and repeatedly, crushing it against its chest until the juicy insides are exposed and can be eaten.

Repopulation Success

Slowly, sea otters have regained their numbers, and today there are about 150,000 of them, about half the number that probably existed 300 years ago. Gradually, they are spreading back to many parts of the coast where they have been extinct for more than a century. It was once thought that sea otters had been eradicated from the California coast, but a few were spotted in 1938, and numbers have lately increased to more than 2,000. Fishermen now say there are too many of them—complaining that the otters eat too many mollusks, crabs, and other valuable sea creatures. Population growth leveled out in the late 1970s, and a small decline may even have taken place since 1998, perhaps indicating that the habitat cannot sustain any more, and sea otters are back at their natural population size.

In an attempt to speed up the recolonization of the sea otter's former haunts, surplus animals have been transported to Washington State, Oregon, and Alaska to repopulate areas along those coasts. Overall the sea otter seems now to have a secure and expanding future in the North Pacific. However, there are new dangers, notably from oil spills near the coast. Detergents used to clean up oil spills are almost as dangerous. Another threat comes from TBT (tri butyl tin), a substance found in the special paint used to prevent barnacles and seaweeds growing on the hulls of boats. The substance also kills other forms of marine life, including some of the main foods of the sea otter. Nevertheless, the sea otter's comeback is one of the best examples of successful international cooperation to secure the conservation of a rare animal.

⤴ The sea otter's dark-brown coloration turns pale cream on the head with age, as with this adult. Sea otters may live for over 20 years and are now a fully protected species.

Common name American badger

Scientific name *Taxidea taxus*

Family	Mustelidae
Order	Carnivora
Size	Length head/body: 16.5–28 in (42–72 cm); tail length: 4–6 in (10–15 cm); height at shoulder: 8–10 in (20–25 cm)
Weight	Male 18–26.5 lb (8–12 kg); female 13–18 lb (6–8 kg)
Key features	Flattened body with short legs, long curved foreclaws, and shovel-like hind claws; gray to yellowish-brown fur with cream belly; sides of face white; dark patches behind ears and on cheeks; white stripe from forehead to nose
Habits	Forages at night; does not hibernate; solitary, except for breeding pairs and family groups
Breeding	Litters of 1–5 young born late March or early April after gestation period of 7 months (including 5.5 months delayed implantation). Weaned at 6 weeks; female sexually mature at 12 months, male at 14 months. May live for 26 years in captivity, 12–14 in the wild
Voice	Normally silent, but occasional yelps
Diet	Mainly burrowing mammals such as pocket gophers and ground squirrels; also birds, reptiles, insects, and occasionally some plants
Habitat	Treeless regions, prairies, meadows, and cold desert areas
Distribution	Parts of Canada, U.S., and Mexico
Status	Population: unknown, perhaps low thousands. Increasing, but still uncommon

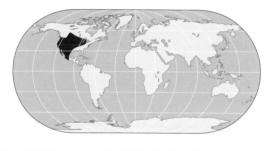

American Badger

Taxidea taxus

The American badger catches most of its food by digging. It can tunnel with amazing speed and will turn over vast amounts of soil in its nightly hunts.

AMERICAN BADGERS HAVE A BODY built for digging. With their powerful claws and partially webbed toes they can move through soil surprisingly quickly. There are stories of badgers digging through asphalt roads. The animals have a third eyelid (a nictating membrane) to protect their eyes from dust. Loose skin on their back and shoulders gives them mobility in tight tunnels.

Multipurpose Dens

Dens are the center of badger life. They are usually simple tunnels with one entrance. Soil excavated in making the den is piled up outside. When badgers are in the den, especially during cold weather, they will sometimes block the entrance with loose soil to help keep warm. The dens used by females to give birth and rear the family are more complex. The side tunnels branch off and rejoin the main thoroughfare, allowing the badgers to pass each other. There are additional side tunnels and chambers, sometimes containing grassy nesting material. Shallow pockets off the main tunnel are dug as latrines and covered in soil. Since they are so extensive, nursing dens have larger piles of soil outside than normal dens. Mounds often contain fur because they are dug in spring when the badger is molting.

American badgers are uniquely adapted for catching underground prey. They dig into burrows to catch pocket gophers, ground squirrels, and many smaller rodents. However, they will also take advantage of whatever other food is available and will eat many small creatures, including snakes, toads, frogs, birds, insects and their grubs, wasps, bees, and worms. In addition they will occasionally eat plant material, too, particularly in the fall, when they take sunflower seeds, corn, and other grains. They also eat carrion and are known to store food in old dens.

Badgers sometimes develop close associations with coyotes, tolerating their presence and even playing with them. The coyotes follow badgers while they are hunting, catching rodents that the badger flushes from burrows. Coyotes help the badgers find new burrowing and hunting areas, sometimes appearing to encourage them with "chase me" play behavior, and sharing the proceeds of their joint hunting efforts.

An American badger emerges from its den. Its powerful shoulders and strong claws make it exceptionally proficient at digging.

Plowing the Land

Badgers are an important part of their habitat because they act as miniplows, literally shaping the land. Their digging loosens the soil and creates patches where different types of plants can grow, increasing the diversity of prairie species. Their holes are often used as ready-made dens by other mammals and as nesting sites by birds such as burrowing owls.

Badgers have few natural enemies, since they are such ferocious fighters. Once they are over a year old and past their vulnerable stage, humans are probably the greatest threat. Many badgers are run over or die in traps set for fur-bearing animals. Others are poisoned by bait put down to control wolves and coyotes. Badger hair has been used to make shaving brushes, but on the whole the animals are not hunted for their fur.

Tolerated

Farmers generally tolerate badgers because they eat large numbers of rodents and will also kill venomous snakes. However, badger burrows can damage crops and are sometimes hazardous to livestock and machinery.

Badgers are one of the few larger mammals that are actually increasing their range in the United States. Because they live in treeless habitats, they benefit from logging and other human activities that open up the land.

Common name
Striped skunk

Scientific name *Mephitis mephitis*

Family Mustelidae

Order Carnivora

Size Length head/body: 12.5–18 in (32–45 cm);
tail length: 7–10 in (17–25 cm); height at
shoulder: 4 in (10 cm). Male larger, but
female has longer tail

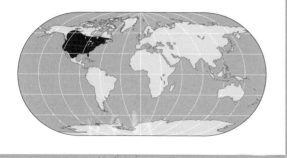

Weight 3–13 lb (1.5–6 kg)

Key features Cat-sized animal, with small head tapering
to a bulbous nose; black coat with forked
white stripes on back; white patch and stripe
on head; long, bushy tail

Habits Mainly active at night and at dusk and dawn;
generally solitary; squirts foul-smelling liquid
when threatened; may swim if necessary

Breeding Three to 9 young born May–June after
gestation period of 62–66 days (including
delayed implantation). Weaned at 6–8 weeks;
sexually mature at 1 year. May live 8–10 years
in captivity, fewer than 3 in the wild

Voice Low growls, grunts, and snarls; also churring
and short squeals; occasional screech or hiss

Diet Mainly insects; also small rodents, rabbits,
birds, eggs, carrion, fruit, vegetables, and
garbage

Habitat Forest or field edges, patches of brush, rocky
outcrops, and wooded ravines; town gardens

Distribution Southern Canada, U.S., and northern
Mexico

Status Population: abundant

Striped Skunk

Mephitis mephitis

Skunks are one of the few mammals to use chemical defense. When threatened, they squirt a repulsive-smelling liquid at their attacker.

THE STRIPED SKUNK IS THE MOST common of the
10 species of skunk. All skunks live in North,
Central, and southern America and do not
occur anywhere else in the world. Skunks are
sufficiently different from other mammals that
they are frequently classified in a family of their
own: the Mephitidae.

The striking black-and-white markings of
striped skunks serve the same purpose as the
stripes on a wasp—they are a warning signal.
After a first unpleasant encounter any animal
that sees another skunk will remember that the
dramatic coloring is associated with a
repugnant smell and will probably make a hasty
retreat. The markings are variable, but usually
consist of a predominantly dark body with a
white blaze on the nose, a white hood, and
white stripes that extend from the hood to the
tail. The tail may be tipped with white, and
there are often white spots on the legs and
around the ears. The stripes on the back vary in
length and width, so that an animal may appear
almost entirely black or white or many
variations in between. There is no difference in
patterning between the sexes, and the
markings do not change with the seasons.

Foul-Smelling Spray

A well-known fact about skunks is that they
stink. Even their Latin name *Mephitis* means
bad smell. The smell is their one effective
defense mechanism, since skunks are not fast
runners, vicious fighters, or clever at hiding.
When a skunk is threatened, it will first give a
warning display by raising its tail and repeatedly
stamping its forefeet on the ground. If the
warning is not heeded, the animal curves its
body into a "C" shape so that it can point its

⊙ *A striped skunk rests
on a log in a field of
flowers. The white
forked stripe that runs
from its head to its
haunches distinguishes
it from other species.*

anus at the target while keeping a close eye on its adversary. The skunk then squirts a yellowish foul-smelling spray from muscular pouches on either side of the anus at the base of the tail.

The spray ("musk") comes through a nipple that can be angled to improve the skunk's aim. It can hit a target at a distance of up to 6.5 feet (2 m). Skunks are able to spray several times in quick succession, and the spray can either be a fine vapor or a more directed shower of droplets. The animals avoid spraying themselves and do not scatter the scent with their tail as many people believe.

The musk contains volatile sulfur compounds, like the smell from bad eggs, which is also sulfur based. Humans can smell it up to 1 mile (0.5 km) downwind; and if it gets into the eyes, it causes extreme pain and even temporary blindness. It takes up to 48 hours for the skunk to replace the musk. The posturing and displaying before the animal sprays is to give their attacker the chance to get away, thereby saving the valuable chemical.

Insects on the Menu

Striped skunks are omnivores. They are opportunistic feeders, eating almost anything that appears vaguely edible. About 70 percent of their diet is made up of insects, such as grasshoppers, crickets, beetles, bees, and wasps. One of their favorite foods is grubs, which they dig up from the soil. Striped skunks will also feed on spiders, snails, earthworms, clams, crayfish, frogs, salamanders, snakes, birds' eggs, small mammals, carrion, berries, nuts, roots, grains, and garbage.

Most food is located by sound or smell, since the skunks' distance vision is poor. In addition, the animals are too slow to chase fast-moving prey. Instead, they hunt like cats, lying in wait or stalking their victim. They catch beetles and grasshoppers by springing on them with their forepaws.

Urban Warriors

Urban areas provide everything that a skunk needs. Human mess means that there is always plenty of garbage to feed on, and where there is garbage, there are usually rats and mice as well. Skunks will also dig up lawns for grubs. They can live in burrows under buildings; and if there are several animals in residence, the smell can be overpowering.

Skunks are unwelcome in towns. Any human that gets too close is likely to be sprayed.

They will use their long front claws to dig for grubs or to tear apart the nests of small mammals, such as mice, rats, moles, and ground squirrels. Striped skunks also break into beehives and will eat the inhabitants without appearing to be affected by the stings. They are known to consume the bee larvae and probably the honey, too. Skunks have a special trick for dealing with poisonous or hairy caterpillars. Before eating them, they roll the prey on the ground with their forepaws, an activity that successfully removes the irritant hairs or toxins in the skin. Striped skunks have also been seen breaking eggs by rolling them between the hind legs with their forepaws until the egg strikes a stone and cracks open.

A Skunk's Way of Life

Striped skunks are not social animals. They come together to breed in spring, but spend the rest of the year alone or in groups made up of mothers and their young. Females occupy a home range of 0.8 to 1.5 square miles (2 to 4 sq. km). Their territory will overlap that of many other females, but they tend to avoid contact with each other. The males travel much farther—about 8 square miles (20 sq. km)—and cover the home ranges of many females, as well as overlapping with other males. The males mate with many females living in their range.

Females prepare maternity dens in March. The young are born in mid-May and are wrinkled, blind, and almost hairless. Even at birth the color patterns are visible as dark patches on the skin. Males do not help rear the kittens and may even attack or kill them, so females defend the maternity dens aggressively.

The kittens are fast-growing: After about three weeks they can assume the defensive posture and squirt scent. They are weaned at six to eight weeks and follow the mother on hunting trips when they are about two months old. When out with their mother, they keep close behind her in a single-file trail. By August they have reached adult size and are able to fend for themselves. They will be able to breed the following spring.

In summer skunks use dens above ground, selecting hollow logs or rock piles. In winter, although they do not hibernate, they rest for

← *Two young striped skunks take in the sounds and smells near their birthplace, a hollow log den.*

sometimes spend the winter in the same burrow as opossums, woodchucks, or cottontail rabbits, but occupy different chambers.

Death by Natural Causes

In captivity skunks may live for 10 years, but in the wild more than 90 percent never reach the age of three. Natural causes of death include starvation in harsh winters, predation, and disease. Skunks are preyed on by great horned owls and some other birds of prey—all of which have a poorly developed sense of smell. Mammalian predators will hunt them too, but only if on the verge of starvation. The smell is enough to deter all but the hungriest hunter.

A common cause of death in skunks is disease. Striped skunks are susceptible to many diseases, including leptospirosis and rabies, which may be passed to humans. Rabid skunks are often very active and aggressive. The virus is present in their saliva, so animals and humans only catch the disease if they are bitten.

Humans are responsible for many skunk deaths, including trapping for the fur trade. Skunks are also killed by vehicles, since they are attracted to roads when searching for carrion.

A striped skunk digs for worms and grubs, its favorite food in a largely insectivorous diet.

long periods in underground burrows, often made by woodchucks or badgers. Some skunks dig their own burrows, but they are not very long or deep. During winter striped skunks will den alone or together in groups of up to 20 animals. Communal denning is more common in the colder, more northerly parts of their range. It seems that skunks will overcome their usual dislike of company to take advantage of the warmth of many bodies. Striped skunks

Skunks at Your Service

Although skunks usually have a bad reputation, they probably do more good than harm. They eat a huge number of agricultural pests, such as armyworms, cutworms, Colorado potato beetles, grasshoppers, beetle grubs, and squash bugs. They are also good at catching mice and rats. Skunks save farmers a lot of money in terms of the amount of pesticides that would be needed if the hungry skunks were not there.

Common name Northern fur seal

Scientific name *Callorhinus ursinus*

Family	Otariidae
Order	Pinnipedia
Size	Length: male up to 6.5 ft (2 m); female 3.7–4.6 ft (1.1–1.4 m)

Weight Male 300–615 lb (136–279 kg); female 66–110 lb (30–50 kg)

Key features Large fur seal; bulls reddish-brown and black, cows pale and more gray

Habits Spends most of the year swimming and diving out at sea; comes ashore to breed in early summer in large colonies

Breeding One young born per year after gestation period of 12 months (including 4 months delayed implantation). Weaned at 3–4 months; females sexually mature at 4 years, males at 6 years but rarely breed before 10 years. May live more than 30 years in captivity, 26 in the wild

Voice Loud bellowing and barking

Diet Mainly fish

Habitat Open sea within 60 miles (100 km) of the coast; comes ashore only to breed

Distribution North Pacific coasts as far south as California; main breeding colonies on Pribilof and Commander Islands

Status Population: about 1 million; IUCN Vulnerable

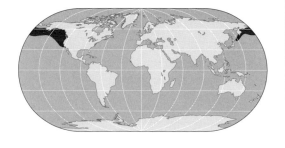

Northern Fur Seal

Callorhinus ursinus

At one time uncontrolled hunting of northern fur seals for their highly prized skins led to a severe reduction in numbers. Today the species is protected, but populations are continuing to decline.

NORTHERN FUR SEALS HAVE recently recolonized Robben Island in the Sea of Okhotsk. About 4,000 now also breed on San Miguel Island off southern California. However, their main breeding sites are on the Pribilof Islands (Alaska) and the Commander Islands off Siberia. Each year the seals migrate to these gloomy, rain-soaked shores, traveling up to 6,200 miles (10,000 km) to breed. The attraction of such places is that the sun rarely shines through the mist and cloud. The lack of sunlight allows the seals to stay ashore for the two months needed to breed. Although thick fur is essential to protect the seals from the cold waters of the Pacific, the animals are known to overheat on land if exposed to sunshine.

Harems of Females
The adult bulls come ashore in early summer and defend a breeding territory. The cows arrive later, and each master bull gathers a harem of up to 100 females for himself. The bulls are four times larger than the cows and have a thick neck. The tough skin is necessary to protect the animals in their fights over dominance and access to females. Pups are born among the boulders on the breeding beaches. A mother can recognize her own pup from all the rest by its unique scent. The mother and pup call loudly to locate each other among the many thousands of other seals. The breeding beaches are therefore a continuous cacophony of bellowing and bleating noises.

Young male fur seals do not normally come ashore for a couple of years after they are born. Even at four or five years old they stand little harvesting for their skins. Pelts were highly prized for their warmth and also for making fashionable fur coats. However, since the fur seal rarely comes ashore except to breed, large numbers were shot at sea. The method was cruel and wasteful because many bodies were never recovered. Once the breeding beaches had been discovered by Russian hunters, the killing took place there. It was an easier operation and more efficient.

Attempts to control exploitation of the seals failed, since it is difficult to protect wildlife living in international waters. The population dwindled to less than 10 percent of the original number, and some colonies died out altogether. But in 1911 the various countries involved agreed to make killing the seals illegal everywhere except on the Pribilof colonies.

For some years after 1911 young, nonbreeding males were targeted in managed culls. Large numbers could be killed with no effect on the overall breeding population because many would never have bred anyway. That was how the fur seal harvest was managed until 1984 (there has been no commercial harvesting since that time). It worked well, enabling an annual harvest of about 40,000 seals to be taken. At the same time, the population was steadily rising to more than 2 million. The northern fur seal became a classic example of successful conservation management involving sustainable harvesting from a wild population.

⊖ *Females, pups, and juveniles of the Pribilof Island stock leave the breeding islands by late November and migrate as far south as southern California and Japan.*

chance of securing a spot on the breeding beaches in the face of the massive beachmaster bulls. Instead, they gather nearby in all-male groups. The beachmasters may be successful for a few years, but then get worn out. Many die before they are more than about 12 years old and are replaced by a few of the youngsters.

At one time there were about 4 million fur seals living in the North Pacific between Kamchatka and Alaska. But before 1911 the numbers were severely reduced by uncontrolled

Mysterious Decline

Although the seals are now protected, their numbers are declining. The main breeding population has halved in fewer than 50 years. About 50,000 northern fur seals drown in fishermen's nets each year. Another serious problem is that fishermen have taken too many fish from the North Pacific, leaving diminished resources for the fish-eating species. It is possible that the food supply may have been reduced to a level that can no longer support the previous numbers of seals.

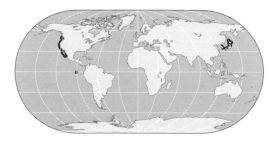

Common name California sea lion

Scientific name Zalophus californianus

Family	Otariidae
Order	Pinnipedia
Size	Length: male 7–8.6 ft (2–2.6 m); female 5–7 ft (1.5–2 m)
	Weight Male 440–880 lb (200–400 kg); female 110–240 lb (50–110 kg)

Key features Typical fur seal with long neck and hind flippers; bulls dark brown, females and young lighter, pups black; adult males have high, domed head, often paler than rest of body

Habits Forms large groups hauled out on rocky shores and on floating jetties in harbors

Breeding Single pup born May–July in California, less seasonal elsewhere, after gestation period of almost 1 year (including 3 months delayed implantation). Weaned at 1 year; females sexually mature at 6–8 years, males at 9 years. May live up to 34 years in captivity, probably fewer in the wild

Voice Loud barking and bellowing

Diet Fish, especially mackerel and anchovy

Habitat Cool seas along rocky coasts

Distribution California to British Columbia (*Z. c. californicus*), around Galápagos Islands (*Z. c. wollebaeki*), and off Japan and Korea (*Z. c. japonicus*)

Status Population: about 200,000 off California; up to 50,000 in Galápagos; IUCN Vulnerable (Galápagos population). Japanese population probably extinct

California Sea Lion

Zalophus californianus

The California sea lion is a familiar inhabitant of zoos and circuses worldwide. It is now also becoming steadily more abundant in its natural home along the western coast of North America.

THE MAJORITY OF CALIFORNIA SEA LIONS occur along the California coast, where they can clearly be seen hauled out on rocks, boat jetties, and floating pontoons. In some places special viewing facilities offer tourists an opportunity to observe these engaging and active creatures basking in the sun and occasionally disputing ownership of the choicest resting places. Up to 50,000 California sea lions also live in the cold waters around the Galápagos Islands about 600 miles (965 km) west of Ecuador. They used to occur on a few islands to the north of Japan, but by 1950 there were only a few dozen individuals left. It is now the case that none have been seen anywhere in the northern Pacific region for almost 50 years. That population of sea lions is probably extinct.

Synchronized Breeding

Male California sea lions range as far north as Vancouver Island, Canada, during the fall and over winter, but females and their young remain closer to the breeding grounds. The breeding sites are mainly on the Channel Islands west of San Francisco, with smaller colonies on islands along the coast of Baja California and around the Gulf of California. The animals gather on their traditional breeding grounds in noisy hordes in early May and remain on the beaches until about July. They select both rocky areas and sandy shores. Ninety percent of all pups are born in June. Highly synchronized production of young is common among the colonial seals and sea lions, and probably helps ensure that predators such as killer whales have so many victims that they cannot kill more than a small fraction. If breeding were spread over a longer time, predators could take far more.

Unlike other colonial pinnipeds, female California sea lions come ashore on the breeding beaches before the males. The bulls arrive later and set out to acquire a selection of females for themselves. They defend an area of 150 square yards (125 sq. m), containing an average of about 16 females. The males stake out their territories along the beach, patrolling laboriously up and down and chasing off rivals for about four weeks, sometimes longer. During that time they do not go to sea to feed, so they lose a lot of weight. Territory holding is stressful and an enormous drain on the body's resources. Nevertheless, many males manage to hold a territory for up to six successive seasons.

Females produce only a single pup each year, born on land. At first the pup has a dark-brown coat, but it molts and is replaced by the juvenile fur at the age of about four to six months. The mother stays close to her offspring for the first week of its life. She will then go back to the sea to feed. Mothers continue to come ashore at intervals for the whole time that they are nursing their pups. Pups begin catching their own food at about five months old, but may continue to take milk also for up to a year. In times of food shortages mothers may even nurse their pups into a second year.

About a month after giving birth, the females come into season and are ready to mate with the males once again. By that time any previous harems and territory systems will have broken down and new ones have to be formed. It is most often the female who chooses her mate, and sometimes groups of females mill around in the shallows seeking the attentions of nearby bulls. Mating takes place in the water or on the lower part of beaches. The breeding behavior in California sea lions

⊕ *In the past, like many other species, the California sea lion was hunted for its skins and oil. Thanks to protective measures, the population has increased from barely 2,000 in 1938 to about 200,000 today.*

is less rigidly controlled than in most other colonial breeding seals, among whom females usually only mate with the bull in whose territory they gave birth.

Closely Packed

Although male sea lions often squabble over who is going to lie in the best resting places on the beaches, they like to pack themselves closely together. In some places they may have less than a square yard (0.8 sq. m) each. When out of water the sea lions are liable to become overheated, especially on sunny days. Then they will wallow in shallow pools among the beach rocks to keep themselves cool.

At sea the sea lions generally make only shallow dives, averaging about 120 feet (37 m) deep. Each dive can be up to 10 minutes long, but usually lasts fewer than three. Sea lions are masters of their watery environment: They can swim at high speeds compared with most other pinnipeds, reaching up to 25 miles per hour (40 km/h), and even faster for short bursts. They can twist and turn abruptly, leap among the waves, and even propel themselves completely out of the water, landing on a flat rock 3 feet (1 m) or more above the surface. None of the true seals can match such agility.

⊕ *Once they have established territories, male California sea lions use ritualized gestures to maintain boundaries. A typical sequence is head shaking and barking as other males approach the boundary (1); followed by stares and intermittent lunges (2); and more head shaking and barking (3).*

Normally, sea lions feed close to the coast, rarely more than about 6 to 8 miles (10 to 13 km) from the shore. Here, they operate in relatively shallow waters up to 250 feet (76 m) deep, catching shoaling fish by swooping among them at high speed. Anchovies and mackerel are their preferred food off the California coast. Some rockfish may be taken near the seabed or among submerged boulders. However, the sea lion is flexible in its feeding habits and may switch to other fish or squid as their relative abundance changes with the seasons. Some years certain species are much more numerous than in others. In "El Niño" years, for example, the sea temperature changes significantly. The fish are scarce and live at greater depths. The sea lions then have to work harder to catch them, and nursing mothers may be unable to produce enough milk for their pup. In those circumstances many young sea lions will die of starvation, and sometimes their mothers, too.

Like many other species of pinniped, the California sea lion was hunted ruthlessly for its skin and oil. By 1938 the population along the California coast had been reduced to barely 2,000 animals. Today California sea lions face few dangers apart from killer whales, hammerheads, and other sharks. Perhaps the biggest threat comes from fishermen's nets, in which sea lions can become entangled and drown. Fishermen also sometimes shoot the sea lions. However, killing is now illegal, since the sea lion is protected throughout its range

⤶ *California sea lions are highly social animals and often congregate in large numbers. They are remarkably graceful in the water, performing agile leaps and turns.*

along the United States coast and off the Galápagos. Protection has enabled a slow recovery in numbers, and by the late 1960s there were about 80,000 along the coast of Mexico and California, rising to 200,000 by 1990. However, numbers sometimes fall back as a result of poor feeding years. Increased abundance has also led to confrontations with fishermen, especially when roaming males visit the rich salmon fisheries off the Canadian coast. Disease hit the Galápagos population badly in the 1970s, leading the IUCN to list that group as Vulnerable. Disease could also strike the California population, especially if the sea lions are weakened as a result of increased numbers and seasonally reduced food supplies.

The Circus "Seal"

California sea lions were, and still are, very popular with the public, and their antics make an interesting show. The animals can clap their front flippers together when told to do so, dive, balance balls on their nose, and retrieve hoops tossed into the water. Their long, flexible necks and mobile hind limbs make balancing acts possible. The true seals cannot perform such tricks, since they are more fully adapted to life in the water and so are less able to perform flexibly on land. Sea lions may take up to a year to learn a new trick. However, they have a long memory and will not forget, even if they have had no practice for several months.

Sea lions are popular performers with the public.

Common name Walrus

Scientific name *Odobenus rosmarus*

Family   Odobenidae

Order   Pinnipedia

Size   Length: male 8.8–11.5 ft (2.7–3.5 m); female 7.4–10.2 ft (2.3–3.1 m)

Weight Male 1,760–3,750 lb (800–1,700 kg); female 880–2,750 lb (400–1,247 kg)

Key features Vast, ponderous seal with bloated appearance; generally pale brown all over; broad, deep snout bears 2 long tusks

Habits   Feeds by diving in shallow seas; spends much time hauled out on shore, generally in company with other walruses

Breeding   Single pup born April–June every 2 years after gestation period of more than 1 year (including 4 months delayed implantation). Weaned at up to 2 years; females sexually mature at 6–7 years, males at 8–10 years. May live over 40 years in the wild, rare in captivity

Voice   Bellowing and grunts; sometimes whistles

Diet   Mollusks, crabs, worms, and invertebrates taken from seabed; occasionally fish

Habitat   Arctic waters along edge of pack ice

Distribution Arctic Ocean: Pacific population along coasts of Siberia and Alaska, Atlantic population mainly around northern Canada, Greenland, and parts of arctic Scandinavia

Status   Population: at least 200,000 (Pacific); about 30–35,000 (Atlantic)

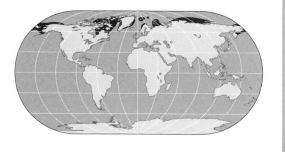

Walrus

Odobenus rosmarus

The vast and unmistakable walrus has suffered from extensive exploitation in the past, but numbers have now largely recovered and are secure.

THE WALRUS HAS A SMALL HEAD, with little piggy eyes and only a tiny fold of skin where the ear would normally be. Its front flippers are almost as broad as they are long. The hind flippers can be folded forward under the belly to help movement on land, just like fur seals and sea lions. The flippers are flexible, rough, and warty on their undersides, to help grip onto smooth ice. A walrus swims at about 4 to 5 miles per hour (7 km/h), although brief bursts of more than 20 miles per hour (35 km/h) are possible. Unlike sea lions (and more like true seals), the walrus propels itself underwater using its hind flippers rather than the front ones. The walrus skull is very solid, but the rest of the skeleton is like that of a fur seal. One speciality is the penis bone, which can be over 2 feet (0.6 m) long— longer than in any other mammal. Not surprisingly, it sometimes gets broken as the massive animals heave their bulky bellies over boulders or lumps of ice.

Conspicuous Tusks

The most distinctive feature of a walrus is its tusks. It was once thought that the walrus used them to pull itself along over the ice. The suggestion is unlikely, although the animals do sometimes use the tusks to heave themselves out of the water and onto the ice. Tusks may also be used to chip away at the edges of breathing holes in the ice in order to keep them from freezing over completely. The conspicuous tusks are probably mainly display features, indicating rank and general superiority, but sometimes the males will use them in fights. Their skin is enormously thickened, especially around the neck, where it can be over 2 inches

(5 cm) thick, with an underlying layer of 4 inches (10 cm) of fat. The fat effectively protects the animals from causing each other serious damage with their tusks, although massive wounds around the shoulders may be caused during territorial fights.

Feeling for Food

The walrus feeds by diving in shallow water, rarely more than 100 feet (30 m) deep. It can stay down for 10 minutes, but most dives are shorter than that. The tusks appear not to be used during feeding and would probably show more signs of wear if they were. Walruses feed on a variety of mollusks that they obtain from the seabed. They have about 450 stiff bristles around their mouths, a shorter and more numerous bunch of whiskers than in any other type of seal. They are embedded in sensitive skin, so that the walrus can gently brush its bristles along the seabed, feeling for solid objects, such as clams. Much feeding has to be guided by touch, especially in murky waters and in winter, when it is dark most of the time. A walrus then uses its massive, fleshy lips to squirt jets of water into the sand at the seabed, thereby exposing the buried clams and other mollusks hidden there. They are then scooped up and may be crushed by the large, flat-topped molar teeth before being swallowed. Usually, the animal eats large clams or seasnails by holding the shells in its lips and sucking out the flesh. It then drops the empty shell and starts to eat another. Doing so saves swallowing lots of shell fragments. Several thousand mollusks may be eaten in a single feeding session. Other seabed invertebrates are found and consumed in a similar way. Prey includes worms, sea cucumbers, and crabs—some of which may be very large, weighing 2 pounds (0.9 kg) or more.

⊖ *The enormous tusks of the walrus are a source of ivory both harder and denser than that of elephants. As a result, the animals have been ruthlessly exploited.*

Over 60 different types of prey have been recorded. Occasionally, fish are eaten, especially flounders and others living on the seabed, although they are not an important source of food. A walrus can consume about 100 pounds (45 kg) of food every day, about 6 percent of its total body weight.

Floating Homes

The walrus lives in shallow water along the continental shelf. It follows the edge of the pack ice, migrating south as the ice extends in winter. As the ice melts in the spring, the walrus travels back again, far to the north. The main American population spends the winter in the Bering Sea, then moves northward around the northern tip of Alaska to spend the summer (July to September) in the Chukchi Sea along the edges of the permanent ice cap.

The annual cycle of migration may involve traveling more than 1,800 miles (3,000 km) each year. The animals prefer to use ice rather than land on beaches. In winter they stay where the ice is relatively thin, but are capable of using their heavy heads to break open ice up to 9 inches (23 cm) thick. The ice holes allow them to breathe and to swim at the surface. In the summer they live among large floating islands of ice and haul out onto them to molt and breed.

Molting takes place first among the younger animals in June and July. Adult males follow, losing their short, bristly hair a month or more before the new coat grows. At that time they may look almost naked, with prominent lumps on the skin around the shoulders. Older males appear to be pink when warm in the sun, but white when they are swimming in cold water. Otherwise, walruses are a cinnamon-brown color all over. Outside the breeding season males and females tend to associate in separate groups. Often the groups may be several thousand strong, with the animals lying packed close together. Yet they are not very sociable and tend to bully each other to establish occupation of the best sites.

During January and February females and

⊕ Tusks signify status in walrus society—the walrus with the largest tusks is generally the dominant animal. This can lead to stabbing duels between dominant animals with similar-sized tusks (1), until one concedes defeat by turning away (2). They occasionally use their tusks to heave themselves out of the sea onto the ice (3) and as head-rests (4).

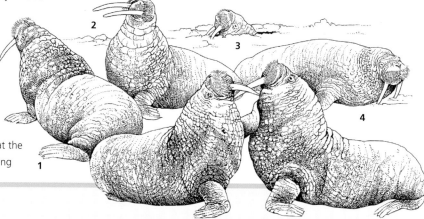

Walrus Ivory

Tusks are present in both sexes of walrus, and very occasionally an animal may have three of them instead of the usual two. Tusks are larger and straighter in the males, and can grow to more than 30 inches (76 cm) long. The longest tusk on record weighed nearly 12 pounds (5.4 kg). The tusks are oval in cross section and have grooves along their length. They are massively extended canine teeth and are unusual in being almost entirely composed of dentine, the material that forms the bulk of our own teeth, with little or no enamel covering. Walrus tusk ivory is highly prized, since it is harder and denser than elephant ivory. The tusks have been used by native people in the Arctic for generations, carved into many types of ornaments, toys, and useful tools. Walrus ivory was a valuable souvenir of Arctic voyages, and extensive hunting of walruses for their ivory led to major declines in populations. The 1989 ban on international trade in African elephant ivory put further pressure on stocks of walruses as a source of substitute ivory, and more than 12,000 are killed every year in Alaska alone. Some scientists believe that an excessive number of animals has been taken in recent years.

younger animals form small parties of about 20, stopping to rest at traditionally used sites. The males compete with each other to stay nearby, each one seeking to take over a whole group to himself. The bulls move around constantly, displaying their tusks, whistling and clicking above water, and making a lot of underwater noise too. Each bull is spaced a safe distance from his nearest major rival. Small males stay out of the way, and only about 10 percent of the biggest and most successful bulls remain with the females. The females slip into the water to select a bull with whom to mate in the dark and icy seas. After mating, the fertilized egg remains dormant inside the female for about four months, then begins to develop normally. Pregnancy lasts a further year, a total of some 16 months in all. Therefore, females cannot produce a pup every year, and pups are born at least two years apart, longer in the case of older mothers.

Caring Mothers

Pups weigh about 140 pounds (60 kg) when they are born. At that time they have pale fur with dark flippers. Mothers look after their young very carefully—with far more devotion than sea lions show. For example, they do not leave their pups for long periods and sometimes will help feed the orphaned offspring of other females. When the pup goes to sea, it may be cradled in its mother's flippers, held close to her chest, or the young walrus can grip the skin of its mother's neck as she swims, towing it along. Like other marine mammals, the walrus produces milk that is 35 percent fat and also

⊕ Two North Hudson Bay walruses, half submerged on an ice floe. The Hudson Bay population has some of the smallest walruses.

Stomach Stones

Pinnipeds of several species, including the walrus, often have stones in their stomachs. The stones are smooth, rounded pebbles, sometimes weighing 1 pound (0.45 kg) or more in total. Some can be as large as tennis balls. Nobody knows why walruses swallow stones, but it is unlikely to be accidental. It is sometimes suggested that padding the stomach with stones helps keep the animals from feeling hungry during long periods without food. More likely, the stones act as ballast and assist in preventing the animals from being too buoyant. Buoyancy makes it harder to dive and stay underwater. Carrying stones in the stomach helps counteract the buoyant effects of blubber and any air remaining in the lungs when the animals dive. It is interesting that crocodiles also carry stones in their stomachs and use them to cancel out the natural buoyancy of their body, helping them stay underwater. Fossil plesiosaurs (large extinct marine reptiles) often have shiny pebbles in the position where their stomach would have been.

rich in proteins. The pups grow well on it, but may still be taking milk from their mother three years after birth, although they are capable of feeding on normal food long before then. They are normally fully weaned at about two years old. Their tusks begin to sprout at about one year. Young females tend to remain in their mother's social group, but males wander off after a year or two. Females become sexually mature at six or seven years old. Males take longer to mature, at eight to 10 years, but will not be able to breed successfully in competition with the big old males until they are fully grown, which may take 15 years.

Traditional Targets

The walrus has played an important part in native folklore and traditions for thousands of years. For centuries it has been hunted for its meat, oil, and ivory and for useful tough hide for making sled covers and tents. Such hunting probably had little effect on the populations, so long as it was carried out using hand weapons and from small boats. Hunters probably killed no more walruses than other predators, such as killer whales and polar bears. However, major exploitation by Europeans began in the 16th century and devastated the populations, particularly in the North Atlantic. In the 19th century more than 10,000 animals were taken each year from parts of the North Pacific population alone. As a result, the walrus became extinct in many of its former haunts, retreating to more remote places where it could not easily be reached. Protection by the Americans in 1909, followed by the Europeans and 50 years later by the Russians, enabled a slow recovery in numbers to a present-day total of about a quarter of a million. Around 10,000 to 15,000 are legally harvested each year by native people. The number may be sustainable if clam fisheries do not also expand, hence taking away vital food supplies for the walrus.

⊛ *A walrus colony on Round Island, Alaska. Despite associating in huge groups, which can be several thousand strong, walruses are not very sociable animals.*

Common name Northern elephant seal

Scientific name *Mirounga angustirostris*

Family Phocidae

Order Pinnipedia

Size Length: male 13–16.5 ft (4–5 m); female 6.5–10 ft (2–3 m)

Weight Male 2–3 tons (1.8–2.7 tonnes); female 1,300–2,000 lb (600–900 kg)

Key features Huge seal with bent, floppy nose; unlike almost all other seals, brown all over

Habits Spends most of its time at sea; occasionally hauls out to rest on rocky islands and beaches

Breeding Single pup born after gestation period of 11 months (including 2–3 months delayed implantation). Weaned at 4 weeks; females sexually mature at about 5 years, males at 8–9 years. May live up to 20 years, but males usually fewer than 12

Voice Bellows and roars

Diet Mostly squid caught in midwater; also some small sharks and slow-moving fish

Habitat Cold coastal waters

Distribution North Pacific coasts of North America from California to northern Mexico

Status Population: about 100,000–150,000

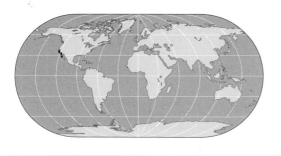

Northern Elephant Seal

Mirounga angustirostris

The northern elephant seal is one of the largest of the world's seals, easily recognized by the drooping, inflatable snout in the adult males. Like most other seals and sea lions, the species has suffered severely from unregulated exploitation in the past.

NORTHERN ELEPHANT SEALS SPEND more time at sea than most other pinnipeds: about 10 months of every year in the case of females. About 80 to 90 percent of their time is spent underwater, which is also not typical of seals in general. Moreover, elephant seals dive very deep, occasionally down to 5,000 feet (1,500 m), where the water pressure is enormous. Such deep dives may last up to two hours, but normally the seals dive for about 20 minutes and only go down about 1,600 feet (500 m). Even so, the average dive time and depth are greater than for most other seals.

There are, in fact, two species of elephant seal. The southern species (*Mirounga leonina*) lives in subantarctic waters, occasionally turning up in scattered locations across the Indian Ocean and South Pacific. Its distribution does not overlap with that of the northern elephant seal, but otherwise the species are similar.

Segregated

Males tend to disperse widely at sea to feed, living alone for most of the year. They travel farther north and farther out into the Pacific than females. Segregating the sexes in such a way avoids competition for food. Elephant seals feed mainly on the squid they find in midwater, often luminescent species that can be seen even in the darkness of deep water. The seals also take small sharks and slow-moving fish, which

⬆ A young male northern elephant seal. Young males are aggressively prevented from breeding by the dominant bulls until they are at least eight years of age.

they find in shallower coastal areas. Elephant seals have little to fear from predators themselves. A fully grown elephant seal is a match for any creature, but youngsters may be taken by great white sharks and killer whales.

Normally, carnivorous mammals have a shorter intestine than herbivores, but the northern elephant seal has a longer intestine than any other mammal. One was measured at 662 feet (201 m), which stretched more than 40 times its own body length and 20 times longer than a typical human gut.

Each year northern elephant seals migrate to their traditional breeding beaches, completing a round-trip migration that totals about 12,000 miles (20,000 km). They gather in December on sandy and pebbly beaches around various islands along the California coast (and at a few sites on the mainland) to set up harems and mate. The bulls compete with each other by showing off their huge bulk. They rear up and bellow fiercely, before bringing their chests crashing to the ground again. They also inflate their big noses to become even more impressive. Sometimes they engage in physical fights, but avoid doing so if possible. During such a contest both males will raise their heads and chests high in the air and wrestle their necks together, attempting to make a downward stab at their opponent with their big canine teeth. Raking the neck with the teeth often opens long gashes, but the skin of an elephant seal's neck is so thick that such wounds are rarely a danger to life, although they do leave impressive scars.

Enforced Fasting

The dominant bulls get to mate with the largest number of females, sometimes as many as 100. Some of the males stay onshore, continuously and without food, until the end of the breeding season in February or March—a total fasting

time of over three months. Meanwhile, the females give birth to their single pup and provide it with milk for up to four weeks. During lactation they too do not go to sea to feed. Instead, they stay close to their pup, feeding it on rich milk derived from their own fat supplies. Converting fat to milk, then giving it to her offspring, means that a typical female may lose nearly half her body weight

⬆ *An elephant seal throws sand over its back. It may be that the sand offers a little protection from the sun.*

before she returns to the sea to feed. Meanwhile, the pup will have grown from a birthweight of about 100 pounds (45 kg) to more than 300 pounds (136 kg).

Toward the end of the suckling period the females will mate, then go to sea a few days later. Development of the fertilized egg is arrested for a couple of months. The embryo then implants itself into the wall of the womb and develops normally through a gestation period of eight to nine months, so that the next pup is born almost exactly 12 months after the last one. At birth northern elephant seals have a black coat, which is soon molted and replaced by silvery hair. After their mothers have

abandoned them, the pups stay ashore without feeding for a month or more. Then they too go to sea and begin to forage for themselves.

The seals haul out on land to molt in early summer—adult males doing so as late as August. Molting in elephant seals is unusual: Instead of losing their hairs one by one as most other mammals do, they shed the outer layer of their skin at the same time. The hair comes off with ragged patches of skin, leaving the animals a sorry sight while their new coat grows into place. The new growth consists only of short, bristly hairs; there is no dense layer of woolly underfur like that found in fur seals and sea lions. Instead, elephant seals (and other members of the true seal family Phocidae) have thick layers of fat (blubber) to insulate their bodies against the cold water in which they live. They might easily get overheated if the summer sunshine was warm during the period they come ashore to molt. However, for much of the summer the California coast is usually bathed in damp fog, screening the animals from the heat of the sun. They come ashore again to breed, but do so in winter, when the weather is bad, and air temperatures are low.

Females may be capable of breeding only three years after they are born, but usually it takes about four or five years to reach sexual maturity. Males take longer to mature and anyway are prevented from breeding by the older and dominant bulls until they are at least eight years old. The stresses of breeding and holding territory on shore while also fasting for three months are so great that few males live more than 12 years.

Targets for Exploitation

Elephant seals were ruthlessly exploited in the 19th century for their blubber, which was melted down to make oil for lubrication and for burning in lamps. A big elephant seal might yield up to 85 gallons (386 l) of fine oil, worth a substantial sum. Over a quarter of a million seals had been killed for the trade by the 1860s. As a result, elephant seals were thought to have become extinct in California waters, but

there seem to have been a few survivors. Some also remained on small islands off Baja California. However, reducing the population to such low numbers meant that present-day elephant seals have all come from relatively few ancestors. That could prove a problem in the future—the population being less resistant to disease, for example. Elephant seals have benefited from protection by Mexico and the United States in the 20th century, and numbers steadily increased to more than 100,000 by the 1990s. Elephant seals have now recolonized almost all the areas from which they were eradicated earlier. Occasional stragglers are now seen as far away as Hawaii and Japan, and north to Vancouver Island and Alaska.

Big Nose!

The huge nose or proboscis of the adult elephant seal has nothing to do with an improved sense of smell. Instead, it is for showing off. The nose grows to full size when the males are mature at an age of about eight years and becomes even larger in the breeding season. It droops, almost like a small elephant's trunk, so that the nostrils point downward, and the end overhangs the mouth by up to 12 inches (30 cm). The strange organ can be inflated by blood pressure and a buildup of air inside, so that it forms a large cushion on top of the animal's snout. The skull is specially shaped and enlarged to accommodate the muscles used to move the proboscis. In its inflated state it acts as a resonance chamber, magnifying the bellows and roars that bull elephant seals make to intimidate their rivals on the breeding beaches.

⊖ *A California male threatens rivals. During the breeding season males compete with each other for mates, and fierce displays of strength are commonplace.*

97

Common name Harbor seal (common seal)

Scientific name *Phoca vitulina*

Family	Phocidae
Order	Pinnipedia
Size	Length: 47–79 in (120–200 cm). Male up to 30% larger than female

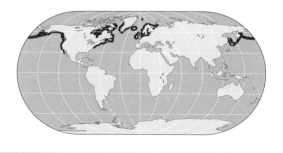

Weight 100–285 lb (45–130 kg)

Key features Typical medium-sized seal; pale-cream coat blotched with mottled pattern of gray, brown, or black; often looks silvery when dry; head small and rounded with short, narrow muzzle

Habits Basks on rocks and sandbanks, usually in small groups; feeds at sea by making short dives

Breeding Single pup born in June and July after gestation period of 10.5–11 months (including 2–3 months delayed implantation). Weaned at 3–6 weeks; females sexually mature at 2 years, males at 5 years. Females may live up to 32 years in the wild, males up to 26 years; probably fewer in captivity

Voice Short barks and grunts, but probably the least vocal of seals

Diet Mainly fish caught on or near the seabed

Habitat Rocky and sandy coasts, estuaries, and sheltered inlets

Distribution Coasts of North Pacific and North Atlantic

Status Population: probably well over half a million. Common and widespread

Harbor Seal

Phoca vitulina

Small groups of harbor seals basking in the sun are a common sight throughout the year along both coasts of the United States and across the shores of northwestern Europe.

HARBOR SEALS SPEND A LOT of their time hauled out to bask in the sunshine. They like to use sandbanks along sheltered coasts, but also sometimes perch themselves on rocky islets as the tide goes out. Normally, they live in small groups, but occasionally several hundred may gather at a good place. Groups may include both sexes. Unlike many seals and sea lions, the species does not form large breeding colonies of male-dominated harems. Indeed, their social structure, if any, is obscure. Also unlike many other seals, the harbor seal is almost silent and only occasionally makes short, quiet barks.

Well-Developed Pups

Mating takes place in the water during the summer months. The fertilized egg remains in a dormant state until the end of the year, when implantation occurs, and true pregnancy begins. Females come ashore to give birth to a single pup in June or July, although along the California coast births occur earlier in the year. Unlike many other seals, baby harbor seals are born with their adult coats already formed and adult dentition already in place. They can swim away on the next high tide if necessary and are capable of making short dives within hours of birth. Nursing occurs in short bouts, lasting only about a minute, every three or four hours. The mother feeds her pup on very rich, fatty milk, enabling it to double its weight within a month or so and be fully independent by the age of six weeks. At weaning the strong bonds between the mother and pup dissolve, and the mother abandons the pup to look after itself, while she begins the process of molting. The process takes place on land, with males following suit a week or two later than females.

Young harbor seals grow fast. The amount of food they need to support their rapidly growing bodies is not always available, and many die at an early stage. About a quarter do not make it to their first birthday, and half never reach breeding age. Some may be overcome by oil slicks or attacked by predators, such as sharks and killer whales. However, apart from human hunters in some places, the species as a whole faces few threats. Some individuals may even live more than 30 years.

In 1988 the North Sea population was hit hard when an outbreak of distemper (a viral disease similar to one occurring in dogs) killed a quarter of the population in a few months. A

Harbor seal pups are fully independent by about six weeks of age. They are born with their adult coats already formed and a full set of adult teeth.

large decline in numbers also occurred along the Alaska coast in the 1980s and 1990s, but its cause is unclear. It is likely that various forms of pollution have weakened immune systems in marine mammals. Sadly, harbor seals live close inshore, where pollution is most concentrated.

Harbor seals feed by making shallow dives normally lasting about eight minutes. However, they can stay underwater for up to half an hour and may dive as deep as 1,400 feet (430 m). Young seals mainly feed on crabs and shrimp, but soon they take to fish, eating a wide range of species. In European waters they often rely mainly on flatfish (like flounders), herring, and cod. Off the Pacific coast they take many spawning fish and migrating salmon. Each seal needs about 5 pounds (2 kg) of fish per day, but will take more if it can. The amount corresponds to about 6 percent of the animal's total body weight. The seals hunt using mainly their eyes, which are large and sensitive in low light conditions, but not so effective out of water. Feeding normally occurs during daylight, but at night and in murky waters the seal's whiskers probably help it locate food.

Staying Close to Home

Some seals undertake long seasonal migrations, but harbor seals tend to stay in the same place all the time. Their daily activity rhythm is dictated by the tide coming in and covering their basking places. Certain individuals may haul out day after day at exactly the same spot, but then disappear for long periods before returning. Occasionally, pups marked at birth may turn up over 600 miles (1,000 km) from where they were born. However, that is unusual and still only a short distance compared with the long and regular migrations undertaken by some pinnipeds.

Harp Seal *Phoca (Pagophilus) groenlandica*

The Arctic harp seal has been the subject of major international controversy due to the harvesting of its attractive white pups for their skins.

Common name Harp seal (saddleback seal)

Scientific name *Phoca (Pagophilus) groenlandica*

Family	Phocidae
Order	Pinnipedia
Size	Length: 5.6–6.4 ft (1.7–2 m)
Weight	250–310 lb (113–141 kg)

Key features Adults light gray with a black face and bold, curved dark marking on the back; pups white

Habits Spends most of year at sea along edges of arctic ice

Breeding Single pup born in February after gestation period of 10 months (including 4 months delayed implantation). Weaned at 10–12 days; sexually mature at about 5.5 years, but often does not breed until about 8 years. May live up to 30 years

Voice Various barks, grunts, and growls; 15 different vocalizations recorded

Diet Shrimp and small fish, especially capelin

Habitat Open seas and edges of ice floes

Distribution Populations based on 3 breeding areas: Newfoundland, Jan Mayen Island, and White Sea, with animals dispersing and going farther north in summer

Status Population: probably at least 3 million. Abundant

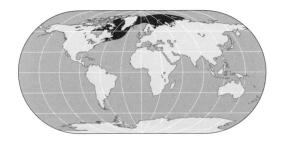

THE HARP SEAL IS SO CALLED because of the characteristic harp-shaped band around its flanks and back. In North America the seal's distinctive markings have given rise to the name "saddleback seal." The harp seal frequents the edges of the arctic ice. It ranges far to the north of Canada in the summer months, often migrating over 3,000 miles (5,000 km) between its breeding grounds and summer feeding areas.

Fishy Diet

Harp seals eat mainly shrimp and small fish, caught near the surface in open waters. However, adults may dive 500 feet (150 m) to feed on herring, cod, and other larger fish. The seals have big, sensitive eyes for locating prey. They make a number of underwater grunts to communicate with each other, but are generally silent above water except at the breeding colonies. They like to live around thick, crumpled, hummocky ice and may travel long distances from the open water by following channels in the ice floes. They can swim fast underwater and move fairly rapidly on the ice.

Harp seals breed in three main areas: Newfoundland, Jan Mayen Island (east of Greenland), and in the White Sea (north of Russia). They assemble in large numbers in late winter (February and March). The newborn seals are nursed two to three times a day on milk that is more than 40 percent fat. As a result, their weight triples in less than two weeks. Pups cry and wail, and their mothers call to them, so breeding colonies are noisy places. After 12 days the pups are abandoned and left to molt. They then go to sea by themselves. The females now mate and disperse. Their fertilized egg undergoes a four-month period of delayed development before full pregnancy begins,

Controversial Cull

Since about 1800 harp seal breeding colonies have been heavily exploited to collect pelts, especially from the white pups. The animals are easily approached across the ice and cannot escape by diving into the sea. Pups are killed in large numbers by clubbing: It has been known for about 180,000 to be killed in a year off the coast of Newfoundland alone. At one time the total population of harp seals may have numbered up to 10 million. However, it became severely reduced, perhaps to less than 30 percent of its potential size. Concern for the species grew, and public opinion turned against culling and the apparently brutal methods used. Campaigns to stop the cull resulted in confrontations on the ice between sealers and animal conservationists. The Canadian government, anxious to protect the traditional source of income for locals, refused access to the breeding areas, even to aircraft and helicopters. The aim was to reduce adverse publicity and to keep opposing factions apart.

Persuaded by public disquiet, European countries banned the import of harp seal skins in 1983. The ban sharply reduced the market, but did not result in the seal hunts being called off. Today about 50,000 pups are still taken each year from the Newfoundland colonies, a similar number from the White Sea, and about 7,000 from Jan Mayen. The culls remain controversial. The effect of a cull on breeding numbers is not seen for at least five years, by which time another quarter of a million pups will have been killed. Yet despite such huge numbers being harvested, the harp seal population remains large. However, many people believe that we should not exert such cruelty on wildlife, especially young animals. On the other hand, the industry provides an income for local people who might otherwise have to move to find work. There is also the problem that the increased harvesting of fish by humans may remove the food supply needed to maintain large populations of seals of all species, the harp seal included.

⊕ The pure white fur of harp seal pups against the white snow is a particularly appealing sight. It is not surprising that feelings run high over the culling of pups.

lasting a further six months. The delay ensures that the pups are born at one-year intervals. Otherwise they would arrive in midwinter, or the males and females would have to find each other again to mate in about May. By mating on the breeding grounds, males and females need only come together once a year.

Females become sexually mature at five years old, (occasionally a little sooner), and males at about the same time. However, in densely crowded populations the animals do not usually breed until they are much older. If numbers become reduced, however, they tend to breed at a younger age.

Common name West Indian manatee (Caribbean manatee)

Scientific name *Trichechus manatus*

Family Trichechidae

Order Sirenia

Size Length: 12–15 ft (3.7–4.6 m)

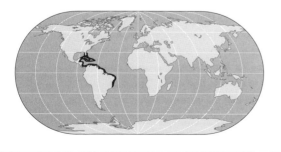

Weight up to 1.4 tons (1.2 tonnes)

Key features Large, sluggish, and slow-moving creature; grayish-brown in color, with paddle-shaped tail and no hind limbs; skin naked with patches of green algae and a few scattered, bristly hairs; blunt-ended head with thick, fleshy lips and small, piggy eyes

Habits Moves slowly, floating and diving in shallow water; often found in small family groups

Breeding Single young born after gestation period of about 1 year, with long intervals between births. Weaned at 2 years; sexually mature at 8 years. May live at least 28 years, probably considerably longer

Voice Normally silent

Diet Aquatic plants, floating and submerged; also grass and other vegetation overhanging from riverbanks

Habitat Estuaries, large rivers, and shallow seas

Distribution Florida, Caribbean, and coastal waters of South America as far as Brazil

Status Population: probably about 10,000–12,000; IUCN Vulnerable; CITES I

West Indian Manatee

Trichechus manatus

Manatees float gently in shallow waters, munching on floating and submerged vegetation. These large but harmless creatures do a useful job keeping waterways free of weeds, but their reward is to be the frequent victims of collisions with motor boats.

WEST INDIAN MANATEES OCCUR around the coastal waters and large rivers of Florida, the West Indies, and the coasts of northern South America. They are sometimes regarded as forming a northern and southern subspecies. The former, based around Florida, probably numbers about 2,000 to 3,000 animals. The southern subspecies—up to 10,000 or so—are spread over the rest of the Caribbean and along the coast of South America as far as Brazil. The animals generally live alone, but sometimes gather in groups of six or more. While they tolerate each other, manatees do not form a true social structure with dominant and subordinate individuals. In fact, the animals may casually join and leave the groups whenever they like. They are probably drawn together by an attractive resource, such as a plentiful supply of food or stretches of unusually warm water.

Signs of Affection

Apart from the close association between mothers and their offspring, there are few interactions between manatees. Nevertheless, they show signs of affection and may indulge in gentle "kissing." There is also evidence that manatees may leave scent marks on logs and submerged rocks. The scent cannot be detected underwater, but may be tasted instead. Such a form of chemical communication is common among land mammals, of course, but unusual in the water. It may indicate that manatees make some sort of social arrangements after all. But there is no sign of territorial behavior, and aggressive interactions of any kind are rare.

A Diet of Greens

Manatees cannot come onto land to forage, since they have no hind limbs. Instead, they dive underwater to feed. As they breathe out, they become less buoyant and sink below the surface. Here they gently paddle around seeking food. They may stay underwater for up to half an hour without needing to breathe. Sometimes they may even take a rest lying on the bottom, apparently asleep. Manatees eat mainly aquatic vegetation, both submerged and floating; in coastal areas they may include mangrove shoots in their diet. Feeding off the bottom is made easier by the West Indian manatee's downwardly directed snout, which enables it to browse underwater beds of sea grass and waterweeds. Altogether, more than 50 plant species have been recorded in the manatee's diet. The animals seem able to consume large amounts of floating water hyacinth, a plant that few other creatures will eat. Water hyacinths are a major pest in tropical rivers, where they clog the waterways and prevent the passage of boats. The manatees' liking for the plant is therefore beneficial to humans, helping maintain open water. Manatees sometimes raise their head well above water to crop overhanging vegetation or nibble at grass along river banks.

Most of the time manatees just float motionlessly at the surface like large logs in the water. With only the top of their back visible, they are difficult to spot and cannot see much themselves. Such behavior is dangerous where

⊕ West Indian manatees move slowly through shallow waters, often traveling in small family groups. As they swim, they feed on aquatic plants that float on the surface or grow under the water.

A West Indian manatee munches on aquatic plants. As the downward-pointing snout suggests, the species feeds mostly on the seabed.

large numbers of fast launches and other boats skim the water. In the coastal waters of Florida and parts of the Caribbean many people have waterside houses and use their boats for fishing and for visiting stores and neighbors. Other boats may be ferrying passengers, towing water skiers, or making local deliveries. Collisions with manatees are frequent and often result in nasty wounds or even death as propellers cut deep into their body. Outside United States waters hunting for their tasty meat is probably the main threat to manatees, but many also drown after getting tangled in big fishing nets.

Breeding Patterns

Breeding can occur at any time of the year. In cooler waters, however, manatees tend to be seasonal breeders, and the calves are mostly born during the summer. A female may mate with several males, not forming a permanent bond with any of them.

Pregnancy lasts for about a year, after which a single calf is born. It stays with its mother for up to two years, sucking milk from one or the other of her two teats, which are located in her "armpits." Within a few weeks of birth the calf also begins to eat plants. By keeping close to its mother, it learns where to find food—a vital skill, especially when seasonal changes in water temperatures may make certain feeding areas unsuitable and force the animals to move on. It will be at least two years before the mother has another offspring. Meanwhile, the calf grows slowly and may not reach breeding condition until it is up to eight years old.

Such a slow rate of reproduction reflects the fact that manatees have few natural enemies and do not need to breed rapidly. Indeed, they might well run out of food sources if they did. However, various unnatural dangers now threaten the animal's placid and stable existence—beside the particular problem of collisions with motorboats. In 1996 over 150 dead and dying manatees were found in Florida waters, apparently poisoned as a result of an unusual bloom of toxic algae. Again, pollution may have been responsible for the huge increase in algal production.

The slow-breeding manatees cannot quickly make good extra losses to their populations, however they are caused. If natural mortality is increased by only a few percent, manatees may easily die out. As a result of injuries and general disturbance, manatees have already declined along the coasts, especially in Florida. They are

Although West Indian manatees lack any cohesive social organization, they will interact with each other using simple gestures, such as "kissing" (1). The mother–calf bond (2), however, is the strongest in the sirenian world. Manatees also communicate via "rubbing posts" (3). They deposit tastes and odors onto prominent objects, where they can be detected by others. Sometimes, manatees relax on their backs on the seabed (4).

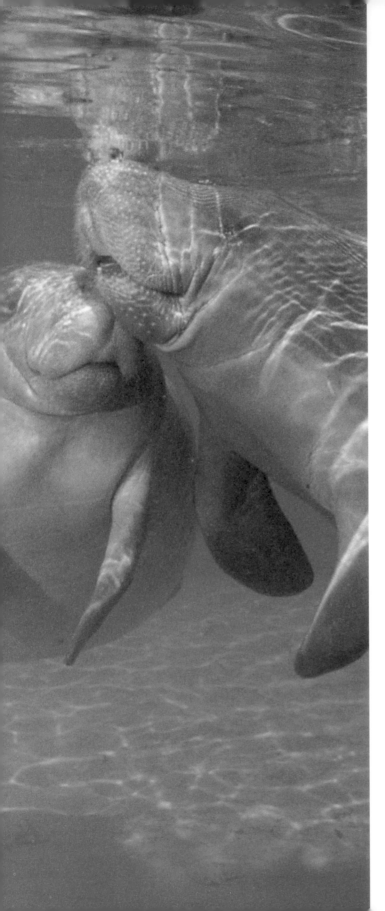

now quite rare in places and may disappear altogether if the activities that cause manatee deaths are not controlled. If the present losses could be reduced, even by a small amount, the manatee population would enjoy a small but steady growth. Boat-free zones and lower speed limits might help protect populations.

Tourist Attraction

Manatees are a popular tourist attraction, especially where they are easily visible drifting in the clear waters fed by springs from the limestone below. However, too many visitors approaching too close may not be beneficial. Local legislation in Florida aims to control the problem. The United States Fish and Wildlife Service has created a special sanctuary at Three Sisters Springs in the Crystal River, where more than 250 manatees spend the winter because the water there is pleasantly warm. Disturbance had been driving them away, but the area will now be off-limits to visitors and boats between November and March.

⊖ A mother and calf. In the 12 to 18 months after birth the young manatee stays close to its mother, learning about feeding areas and annual migration routes.

A Warm Welcome

Manatees do not like water that is cooler than about 68°F (20°C). In summer Florida's manatees disperse widely along coasts and rivers. In cooler periods, particularly in winter, they make local migrations and often gather together in places where power plants discharge warm water into the sea, such as at Cape Canaveral, Fort Myers, and in Tampa Bay. Farther south in the Caribbean the water remains warm most of the time. Here manatees tend to move around less according to season, congregating to feed rather than to enjoy warm water.

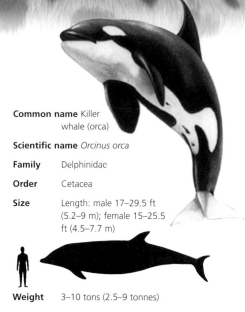

Common name Killer whale (orca)

Scientific name *Orcinus orca*

Family Delphinidae

Order Cetacea

Size Length: male 17–29.5 ft (5.2–9 m); female 15–25.5 ft (4.5–7.7 m)

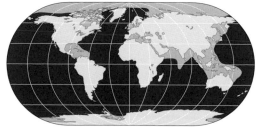

Weight 3–10 tons (2.5–9 tonnes)

Key features Striking black-and-white markings; body mainly black with white patch behind eye, white cheeks and belly, and gray saddle patch; head rounded with no obvious snout; tall, triangular dorsal fin, up to 6 ft (1.8 m) high in male; broad, rounded flippers; tail black on top, white on underside

Habits Social, living in a tight-knit family group or "pod"; fast, active swimmer; acrobatic at the surface, will breach, spy-hop, and tail slap

Breeding Single calf born about every 8 years after gestation period of 17 months. Weaned at 14–18 months; sexually mature at 12–16 years. Males may live 35–60 years, females up to 90 years in the wild; rarely survives more than a few years in captivity

Voice Varied, including complex, often pulsed, calls

Diet Ssmall fish and squid to seals, turtles, seabirds, and even other whales

Habitat Open sea to coastal waters; estuaries; often around ice floes in polar waters

Distribution Every ocean in the world, from polar regions to equator

Status Population: 100,000; IUCN Lower Risk: conservation dependent; CITES II. Widespread and quite numerous

Killer Whale/ Orca

Orcinus orca

The largest member of the dolphin family, the killer whale (or orca) is the top sea predator. Hunting in groups, orcas will even attack giant blue whales.

KILLER WHALES ARE ONE OF the most recognizable cetaceans, familiar from their antics in captivity and in the *Free Willy* movies. They are very large, heavily built dolphins with characteristic black-and-white markings. Their muscular bodies make them the fastest mammal in the sea, with sprints recorded at 35 miles per hour (56 km/h)—almost as fast as a racehorse.

From a distance the most recognizable feature of a killer whale is the tall, triangular dorsal fin. In a mature male it can be up to 6 feet (1.8 m) tall—the height of a man. In females the dorsal fin is only half as tall and has a more curved shape. The flippers are also large, especially in males. Killer whales have 20 to 26 sharp teeth in both the top and bottom jaws. The teeth are pointed, conical, and each is up to 2 inches (5 cm) long. When the jaws close, they interlock perfectly, clamping fish and other prey in a vicelike grip.

Not So Deadly

Killer whales are formidable hunters and gained their name from 18th-century whalers who saw them attack other whales and believed that they would also eat humans. This belief lasted until the 1960s, when people began to study whales more closely. Despite their reputation, there are no records of humans ever being killed by a killer whale, so some people prefer the more kindly name of "orca."

Killer whales are one of the most widely distributed animals in the world. They live in every ocean and have adapted to both the icy conditions of the Antarctic and the warm, equatorial seas. However, individuals do not appear to migrate between them as some

Killer whales are the top sea predators. They will catch and eat almost any type of prey—from seals and turtles to other whales. However, they tend to feed on locally abundant resources and this can affect hunting techniques and even body size in different parts of their range.

species of whale do. They are more common in cold, polar waters. One of the best places to see orcas in the wild is the waters of the Juan de Fuca Strait and San Juan Islands between Washington State and Vancouver Island, British Columbia, Canada.

Killer whales live in social groups known as "pods." These consist of up to 50 animals, usually one mature male, several mature females, and young of both sexes. They are stable, tightly knit groups, with animals staying with the same pod for their whole life.

Group Dialects

Animals in a pod communicate with complex calls, some of which may serve to identify the group to other pods. Each pod has a distinct dialect, using a characteristic pattern of repetitive sounds. The dialects are so distinctive that by listening to the calls, even a human researcher can tell which pod an animal is from.

Three types of killer whale have been identified, each with different social habits. There are residents, transients (or wanderers), and offshore animals. Residents stay in large family pods, usually of five to 25 animals, and have a relatively small home range. They tend to be very noisy, communicating to each other with frequent calls. They hunt using echolocation and feed mainly on fish and squid. Transient orcas live in smaller pods with one to seven animals. They roam over a wide area and are quieter, using stealth to hunt sea mammals such as seals, sea lions, and other dolphins. Offshore orcas spend most of their time in the open sea, much farther away from the coast, and probably eat mainly fish. They seem to stay in large groups of 25 or more, communicating with each other frequently and noisily.

Killer whales will catch and eat almost any type of prey. They have been recorded eating over 100 species of animal, more than any other cetacean. As well as fish and squid, they will eat seals, dugongs, turtles, penguins, gulls, and even other whales more than 10 times their size. When they attack large whales, they tend to bite pieces off their lips and tongue.

They hunt cooperatively in a team, like a wolf pack, and use different techniques to catch different prey. They will herd salmon by making noisy calls and slapping their flippers on the surface, trapping the fish in a tight, frightened bunch before lunging in to eat them. Seals and sea lions are one of the killer whale's favorite foods, and the whales go to great lengths to catch them. Prey animals are not even safe out of the water: In parts of Argentina and on the Crozier Islands in the Indian Ocean killer whales launch themselves onto the beach to seize baby sea lions resting there. Then, using their front flippers, they wriggle back into the sea to eat their victim. Killer whales in antarctic waters have been known to hunt seals by tipping them off ice floes. A researcher watched a group of whales, their heads poking out of the water, scanning the ice for seals. When they spotted one, they watched it carefully for a few minutes. First, the whole group swam rapidly toward it, then all together dived under the ice. Their dive caused a big wave to wash over the loose ice floe, tipping it at such a sharp angle that the seal slid off into the sea where the whales could catch it. Such an example of cooperative behavior suggests that killer whales are very intelligent creatures, capable of problem solving and coordinating their actions as a group for the benefit of all.

⬆ *A killer whale off the coast of Patagonia, Argentina, snatches a sea lion for its meal.*

➡ *Killer whales are social animals that live in tightly knit groups known as "pods."*

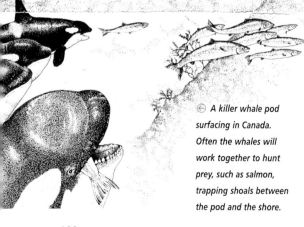

⬅ *A killer whale pod surfacing in Canada. Often the whales will work together to hunt prey, such as salmon, trapping shoals between the pod and the shore.*

Killer whales have no natural predators; but as with many other species of whale, human interference causes many deaths. Some killer whales are hunted, although less so now than in the past. In the years between 1938 and 1981 Japanese, Norwegian, and Russian whalers killed 5,537 killer whales. A few orcas have also been caught for exhibition in major aquaria—over 150 since the 1960s. Once in captivity, most killer whales only live for a few years, but offer a fascinating and educational spectacle for millions of visitors each year.

Salmon Stealing

Killer whales also come into conflict with fishermen, who will sometimes kill them if they believe that they are taking valuable fish. In Alaska killer whales have learned to steal salmon directly from fishing hooks. Fish farms can also cause problems. Intensively farmed fish are a source of diseases, and pesticides used to control fish lice can get into the food chain.

Studying Killer Whales

Orcas in the waters around Vancouver Island in British Columbia, Canada, have been studied for many years. Individual animals can be identified by the different shapes and sizes of their dorsal fins. Some have distinctive nicks and scars on their bodies. The shape of the gray saddle marking often varies, enabling individuals to be recognized. By following particular animals from birth, recording their family histories, habits, and interactions with other orcas, researchers are learning more and more about the complex lives of these previously misunderstood creatures.

Other threats include boat traffic and underwater noise from engines. The killer whale is also particularly vulnerable to pollution, which affects its prey and can lead to the accumulation of dangerous substances in the whales' own body tissues.

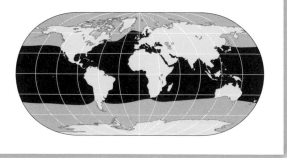

Common name Common dolphin

Scientific name *Delphinus delphis*

Family	Delphinidae
Order	Cetacea
Size	Length: 5–6.5 ft (1.5–2 m). Male generally larger than female

Weight 155–297 lb (70–135 kg)

Key features Fairly large dolphin with long, slender beak; distinctive "hourglass" pattern on flanks made by a wide yellow band from head that closes to a point below dorsal fin, opening up again as a gray band and darkening toward tail; dark back, flippers, and fins; creamy-white underside with black "chinstrap" extending from lower jaw to flippers

Habits Active, acrobatic, and noisy dolphin; fast swimming; sociable: normally found in groups of usually a few dozen animals

Breeding Single calf born every 1–2 years after gestation period of 10 months. Weaned at 12–18 months; sexually mature at 5–6 years. May live about 25 years in the wild, rarely kept in captivity

Voice Pulsed whistles, clicks, and high-pitched squeaks

Diet Squid and shoaling fish

Habitat Waters with a surface temperature of above 50°F (10°C), usually more than 600 ft (180 m) deep

Distribution Widespread in warm-temperate, tropical, and subtropical waters

Status Population: abundant, many millions; CITES II. Intensive hunting in Black Sea has reduced local population. Pacific population has suffered as a result of being accidentally caught by tuna fisheries

Common Dolphin

Delphinus delphis

The common dolphin is a sociable, boisterous, and noisy cetacean that lives in warm waters throughout the world. It has recently been split into two different species: the long-beaked and short-beaked common dolphins.

THE NAME *DELPHINUS* IS THE Latin and Greek word for dolphin, and the common dolphin is the species that the ancient Greeks and other early civilizations were most familiar with. Common dolphins were often depicted on their pottery and wall paintings.

The common dolphin lives in warm seas all over the world. On the American side of the Pacific they can be seen from British Columbia down to central Chile, and in the Atlantic from the northeastern United States to Argentina.

Elaborate Markings

The common dolphin is fairly large, with a streamlined body and a long, slender snout (often termed a "beak.") The markings vary between different populations of common dolphin, but are generally much more elaborate than on any other species of whale or dolphin. The dark cape (which can be a brown, black, gray, or purplish color) is a characteristic feature. It ends in a "V"-shape below the fin, with a broad, yellowish band from the eye to the middle of the body. The belly is white and may have another yellowish stripe below the main yellow band. A gray band along the flanks joins the yellow one in a point. Together, the colors make a distinctive "hourglass" pattern. The tail flukes are black, and there is a black "chinstrap" stripe from the flippers to the lower jaw. There may also be another black stripe from the tail area into the side of the belly.

Common dolphins are highly sociable animals that travel, feed, and sleep in groups. Group size depends on the season and whether it is day or night. In most areas groups of 10 to 500 animals are usual, but occasionally— particularly in the eastern tropical Pacific— groups can be over 2,000-strong. They have been known to hunt cooperatively, working as a group to herd fish into a tightly packed shoal where they are easy to grab by the mouthful. When the dolphins are frightened, the group bunches tightly together for defense.

Acrobatic Displays

The dolphins are very acrobatic. They can often be seen jumping into the air, somersaulting, lobtailing, and flipper-slapping. They frequently ride on the bow waves of boats or even larger whales. They are fast swimmers, reaching speeds of up to 27 miles per hour (43 km/h). Groups of fast-moving dolphins tend to arch out of the water to breathe at the same time, a behavior known as "porpoising." They often cannot be seen for the mass of foam they create. Common dolphins usually only take short dives of about 10 seconds to two minutes, with a maximum of eight minutes. They are one of the noisiest of the small whales, using clicks for echolocation and squeals and high-pitched whistles for communication. They are so loud that people on boats nearby can hear them.

⊕ *Common dolphins are highly sociable animals that travel, feed, and sleep in groups. In the Pacific groups can number up to 2,000 animals.*

The wide variations between different populations of common dolphin have led some biologists to try to divide them into distinct species. It is now thought that there are probably two species of common dolphin; the short-beaked (*Delphinus delphis*) and the long-beaked common dolphin (*D. capensis*). The long-beaked dolphins have a slightly longer snout and more muted colors than the short-beaked types, with less contrast between the black and yellow-white markings. In addition, they have a flatter forehead that meets the beak at a shallower angle than in short-beaked dolphins. The two species can also be told apart biochemically by differences in their DNA (genetic molecular structure).

Fishing Net Deaths

Common dolphins are sometimes caught illegally by fishermen in Japan, South America, and the Azores. They are also one of the most frequent species caught accidentally in fishing nets. Large nets that were used to catch tuna trapped many dolphins, killing thousands every year. The animals could not surface to breathe and consequently drowned. Methods of making the nets more conspicuous to dolphins and built-in escape routes for the trapped animals, have now been introduced. Such measures should help reduce the death toll.

111

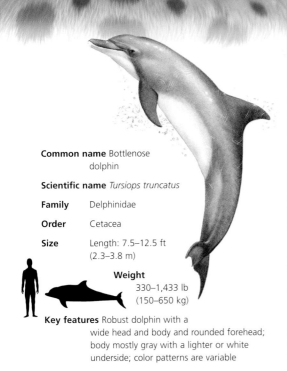

Common name Bottlenose
dolphin

Scientific name *Tursiops truncatus*

Family   Delphinidae

Order   Cetacea

Size   Length: 7.5–12.5 ft
(2.3–3.8 m)

Weight
330–1,433 lb
(150–650 kg)

Key features Robust dolphin with a
wide head and body and rounded forehead;
body mostly gray with a lighter or white
underside; color patterns are variable

Habits   Active, social dolphin usually seen in groups

Breeding   Single calf born every 4–5 years after
gestation period of 1 year. Weaned at 4–5
years; females sexually mature at 5–12 years,
males at 10–12 years. May live up to 50 years
in the wild, fewer in captivity

Voice   High-pitched whistles and clicks

Diet   Large variety of food, including fish, squid,
octopus, cuttlefish, and mollusks

Habitat   Wide range of habitats from open water to
harbors, bays, lagoons, estuaries, and rocky
reefs

Distribution Widespread in temperate and tropical
waters

Status   Population: unknown, perhaps hundreds of
thousands; CITES II. A common species,
especially in particular areas

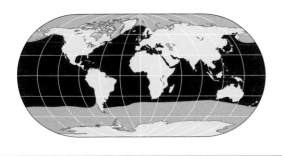

Bottlenose Dolphin

Tursiops truncatus

Bottlenose dolphins are highly social, intelligent animals. They form close, cooperative "friendships" with each other, and some individuals will even seek out human contact.

BOTTLENOSE DOLPHINS ARE THE MOST familiar species of small cetaceans, made famous by the televison series *Flipper*. They are the best studied of all whales, partly because their fondness for coastal waters makes them easy to observe, and also because they adapt to captivity better than other species. They are the dolphins most frequently seen along the shores of the United States.

Variable Body Shape

Bottlenoses are robust dolphins, with a short, wide snout (or "beak"). Their size and pattern vary widely depending on where in the world the dolphins live, and whether they come from coastal or offshore waters. In the northwestern Atlantic coastal dolphins tend to be smaller and slimmer than their offshore counterparts. Body size also depends on the temperature of the water: Dolphins living in colder seas need more blubber to keep warm. Because five of their seven neck bones are not fused rigidly together as in other whales, they have more mobility in their necks than most other cetaceans and thus can nod and turn their heads.

Bottlenose dolphins live in a wide range of temperate and tropical seas. Along the western Atlantic coast they can be seen from New Jersey to the Caribbean and Panama. On the other side of the United States they are found along the coast from Panama to southern California. Since color, markings, and healed scars make each individual different, it has been possible to keep a photographic catalog of the animals. Studies show that the dolphins living near the coast return to the same area year after year.

Depending on their habitat, most dolphins do not need to dive very deep to catch their food. They regularly go down to depths of between 10 and 150 feet (3 and 46 m), holding their breath for eight to 10 minutes.

The bottlenose dolphins' diet is very varied. They are second only to killer whales in the number of species they eat. They will take fish including sea trout, anchovies, herring, and cod, and invertebrates such as squid, octopus, and large shrimp. Under experimental conditions blindfolded dolphins are able to find fish, even small ones, by using their underwater echolocation system (or sonar). They will adapt their diet to the conditions where they live, and some groups have developed special feeding techniques. In the Gulf of Mexico dolphins sometimes catch large fish by flicking them out of the water with their tails. The stunned fish can then easily be picked up from the surface.

Social Structure

Bottlenose dolphins are nearly always found in groups. In coastal waters group size is usually fewer than 20, but offshore gatherings of hundreds are sometimes seen. Dolphins have a loose social structure, with individuals coming together then separating, and joining up with other dolphins. There are three main types of groups: The first is mother and calf pairs or groups of mothers with their most recent offspring. Such associations tend to stay in the parts of their home range where food is most plentiful. Young animals will stay with their mother for up to five years. Once they leave their mother, young dolphins will stick together in gangs. Such subadult groups may be mixed—with males and females—or single sex. As they get older, the animals range farther afield and spend time in smaller and smaller groups. Males eventually form strong, long-term alliances with one or two other males. The groups of males will move between female groups in their range. Females, however, tend to form looser alliances that are more flexible. Females will help each other give birth to, raise,

⊕ Bottlenose dolphins live in a wide range of seas, from tropical to temperate. They are often seen in captivity, and are therefore one of the more familiar cetaceans.

Bottlenose dolphins are very active, acrobatic animals. They will leap up to 30 feet (9 m) into the air, ride the bow wave of boats, and bodysurf in breaking waves. Both adults and young are sometimes seen "playing" with objects such as seaweed, coral, or even jellyfish. They carry the objects, throw them around, and use them to invite other animals to play. Bottlenose dolphins also seem to enjoy blowing bubbles—dolphins in captivity have been seen blowing perfect halos by trapping a bubble in a vortex of water (a bit like a smoke ring).

and teach young. They will even protect each other from any unwanted male advances.

Social bonds between individual dolphins appear to be strong. Studies show that certain animals prefer the company of "friends" and recognize each other after long periods of separation. Captive dolphins have even been trained to communicate instructions to each other. Physical contact is frequent. Dolphins will stroke and caress each other, and contact of a sexual nature, including copulation, is used to reinforce bonding. Dolphin society can also be very aggressive: The animals will vocalize angrily with clicks, squawks, and pops. Some show their dominance by raking others with their teeth, frequently leaving extensive scars.

Bottlenose dolphins are very curious and often swim with boats and bathers. There are many cases of dolphins staying in one area and deliberately seeking out human contact. Fungi, a bottlenose dolphin that has lived in Dingle Bay, Ireland, since 1984, is one of the most famous. Extrafriendly dolphins are usually lone males that appear to have no natural social group of their own.

Dolphins to the Rescue

There are many stories of dolphins helping people. Dolphins have come to the rescue of drowning sailors, lifting them to the surface and either helping them reach the shore or keeping them afloat until help arrives. They have also been known to protect people from shark attacks. In July 1996 a man was swimming in the Red Sea with five bottlenose dolphins when a shark attacked him. Three of the dolphins surrounded the swimmer and by slapping their fins and tail flukes prevented the shark from attacking again until help arrived. Some people have claimed that contact with dolphins can trigger the healing process in humans. Swimming with dolphins has been used as therapy for many conditions, including depression, anxiety, and cancer. It has also helped stimulate learning in people with disabilities. In some parts of the world bottlenose dolphins cooperate with fishermen. In Santa Catarina, Brazil, dolphins help local fishermen by driving the fish toward their nets. The United States Navy has trained dolphins to help with underwater mine clearance.

Dolphins are highly intelligent. They are one of the few animals that have been shown to be self-aware in that they can recognize themselves in a mirror. The only other animals capable of doing so are humans and great apes. Trained dolphins are also able to respond to at least 20 different commands—a better repertoire than most sheepdogs.

Natural predators of bottlenose dolphins include tiger and bull sharks, and occasionally killer whales (orcas). Humans, however, cause their biggest problems. There have been cases of mass deaths of dolphins from viral diseases. Often the victims have been heavily contaminated with pollutants that may have damaged their immune systems. Bottlenose dolphins are still caught in Japan for their meat. They are also captured for use in displays. Bottlenoses are the species most commonly kept in captivity. Between 1860 and 1983 more than 2,700 were caught for exhibition in dolphinaria and zoos, with over 1,500 taken from United States waters. Dolphins are now protected in the United States by the Marine Mammal Protection Act of 1972.

⊖ *Gregarious and intelligent, bottlenose dolphins form strong social ties. Individuals often appear to seek the company of certain special "friends."*

Cooperative Fishing

As would be expected of such intensely social animals, dolphins often hunt cooperatively. In South Carolina and Baja California dolphins have learned an extraordinary trick to catch fish. They work together to herd a shoal of fish close to the shore. Then swimming in a tight, fast line, they sweep them onto the beach. The dolphins then roll onto the beach, out of the water, to grab the stranded fish.

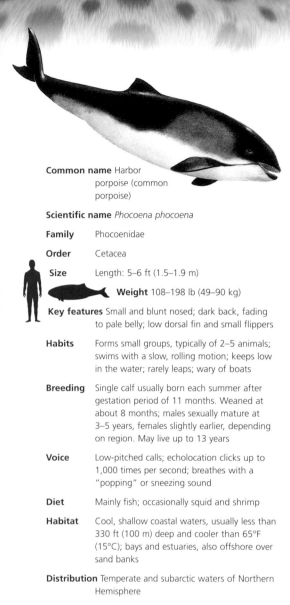

Common name Harbor porpoise (common porpoise)

Scientific name *Phocoena phocoena*

Family Phocoenidae

Order Cetacea

Size Length: 5–6 ft (1.5–1.9 m)

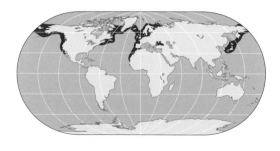

Weight 108–198 lb (49–90 kg)

Key features Small and blunt nosed; dark back, fading to pale belly; low dorsal fin and small flippers

Habits Forms small groups, typically of 2–5 animals; swims with a slow, rolling motion; keeps low in the water; rarely leaps; wary of boats

Breeding Single calf usually born each summer after gestation period of 11 months. Weaned at about 8 months; males sexually mature at 3–5 years, females slightly earlier, depending on region. May live up to 13 years

Voice Low-pitched calls; echolocation clicks up to 1,000 times per second; breathes with a "popping" or sneezing sound

Diet Mainly fish; occasionally squid and shrimp

Habitat Cool, shallow coastal waters, usually less than 330 ft (100 m) deep and cooler than 65°F (15°C); bays and estuaries, also offshore over sand banks

Distribution Temperate and subarctic waters of Northern Hemisphere

Status Population: 200,000–300,000; IUCN Vulnerable; CITES II. Decreasing, mainly due to accidental capture in fishing nets, but also problems with disturbance, food supply, and disease

Harbor Porpoise

Phocoena phocoena

The harbor porpoise is a placid, shy creature that tends to stay unobtrusively underwater. It is the commonest and most well-studied member of the porpoise family, glimpsed fleetingly around coastlines in estuaries and bays.

THERE ARE SIX SPECIES IN THE porpoise family. Most live in coastal waters, and all are small—less than 6.5 feet (2 m) long. Dolphins are often called "porpoises," but unlike dolphins, the true porpoises all have a blunt-ended face with no projecting snout or "beak."

The "Pig Fish"

The Romans used to call porpoises "pig fish," or *porcus piscus* in Latin, which gives us the name porpoise. Porpoises have flattened, spade-shaped teeth (rather like human front teeth), instead of the usual conical, pointed teeth of most other toothed whales. These teeth are good for grabbing and gripping the large, smooth fish that make up most of the porpoises' diet.

The harbor, or common, porpoise is the most frequently seen porpoise in the wild. It comes into bays and estuaries, and sometimes swims quite long distances up rivers. Most sightings are within 6 miles (10 km) of land, while many other species of small cetaceans are found only well out to sea. In the United States the harbor porpoise can be seen along the Atlantic coast south to the Carolinas and occasionally into Florida; on the Pacific coast they occur from Los Angeles to Alaska.

It is easy to identify a harbor porpoise in the wild because of its habitat, small size, rounded face, and small, blunt-tipped dorsal fin. The coloring is dark gray on the back, fading

⊕ Harbor porpoises are rather shy creatures. They do not show some of the more exhibitionist traits common to many of the dolphins and larger whales, such as breaching, bodysurfing, and flipper-slapping. It is actually fairly rare to see much of the porpoise above the water.

down the flanks to white on the belly. Harbor porpoises have a black chin and lips and a mouth that curves up slightly so they look as if they are smiling. Their breathing is also distinctive—a sudden outburst that sounds like a pop or sneeze.

Harbor porpoises usually swim in small groups of two to five, but never form large schools like some of the dolphins. Groups sometimes come together to feed when there is a large shoal of fish or other prey, suggesting they can communicate over long distances. However, there is no evidence for the strong social structures that exist in some dolphin species. They also do not swim in formation like some dolphins do, with all animals coming up for air at the same time. Harbor porpoise groups tend to appear as more of a disorganized "rabble," with individuals crossing each other's paths and making sudden spurts and charges. They are not the most exciting cetaceans to watch because they are timid and tend to show very little of themselves above water. They never leap high or ride the bow waves of ships or perform any of the other interesting behaviors so characteristic of

dolphins and many of the larger whales. Usually they just swim slowly, surfacing with a slow roll every 10 to 20 seconds, making a "pop" sound when they breathe.

Decreasing Numbers

Harbor porpoises were once very common, but their numbers have decreased rapidly owing to human activities. The coastal waters where they live are often very heavily polluted, and many of their habitats are lost as natural coastlines are taken over for industry or shipping. Boat traffic is another hazard for such shy creatures, and there have also been cases of large numbers dying from disease.

Harbor porpoises are still hunted in some parts of the world, with the largest catches around Iceland. Many thousands also drown in fishing nets—particularly in gillnets, which hang vertically in the water anchored to the seafloor. They are used to catch low-swimming fish such as cod and flounder; but since porpoises also tend to feed near the seafloor, they often become trapped. The fishing industry is trying to find ways to reduce the number of dolphins caught in gillnets. One method is to attach devices called "pingers" that make underwater noises to warn the animals to keep away.

Beluga

Delphinapterus leucas

The beluga is unique among whales owing to its ability to produce several facial expressions. It can alter the shape of its forehead and lips, often appearing to smile, frown, or whistle.

Common name Beluga (white whale)

Scientific name *Delphinapterus leucas*

Family	Monodontidae
Order	Cetacea
Size	Length: 10–16 ft (3–5 m). Male larger than female

Weight	1,100–3,300 lb (500–1,500 kg)
Key features	Stocky, white-colored whale; no dorsal fin; head small and rounded; flippers broad, short, paddle shaped, and highly mobile; tail fluke frequently asymmetrical
Habits	Social animals, rarely seen alone; masculine groups of 3–15, nursery groups of mature females and several young of various ages; during migrations congregations of hundreds or even thousands may be seen
Breeding	One calf born every 3 years after gestation period of 14–14.5 months. Weaned at 20–24 months; females sexually mature at 5 years, males at 8 years. May live 30–40 years in the wild, some have been known to live to 50 years; does not survive so long in captivity
Voice	Trills, moos, clicks, squeaks, and twitters; sometimes called "sea canary"
Diet	Mostly bottom feeders, eating fish, crustaceans, worms, and mollusks
Habitat	Coastal and offshore in cold waters, usually near ice; shallow waters, rivers, and estuaries
Distribution	Coasts of arctic regions of North America, Greenland, northern Russia, and Svalbard
Status	Population: about 100,000; IUCN Vulnerable; CITES II

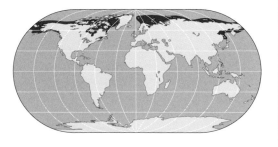

For most of the year adult belugas are pure white, only becoming tinged with yellow for a period in the summer before the seasonal molt. Their coloration explains their common name of white whale. Calves are dark gray when first born, but the color fades during their immature years until they reach about four years old, when they become white. The lightening in color is due to a reduction in the dark pigment (called melanin) in the skin. In summer belugas shed the surface layers of their old, slightly yellow skin to reveal the new, gleaming-white skin underneath. At that time the animals like to congregate in estuaries; but if that is not possible, they head for melting glacier fronts instead. Estuaries and glaciers are both sources of fresh water, which may quicken the shedding process. The belugas also rub themselves on the seabed to help themselves molt.

Layers of Fat

The beluga is quite a small and rotund whale with short, wide flippers that curl up at the tips. There is no dorsal fin—accounting for its scientific name *Delphinapterus*, meaning "dolphin without a wing." However, there is a short, raised ridge where the fin would normally be found. The beluga has a very thick layers of blubber to keep it well insulated in the cold arctic seas. It is so fat that the head looks too small for the body. Unlike most other whales, belugas have a very mobile neck that allows them to nod and turn their head. They are inquisitive animals and use their neck to look around when they lift their heads out of the water. Another feature of the beluga is its ability to break through solid ice up to four inches (10 cm) thick by ramming it from

The belugas' deathly white skin and undulating swimming motion makes them look like underwater ghosts. For a period in the summer belugas become tinged with yellow before the seasonal molt occurs.

underneath, using the firm melonlike structure on its head. The impact creates breathing holes in packed ice. However, belugas cannot stay in waters that are covered by very thick ice, since they would not be able to break through it to breathe the air they need.

Ghosts of the Deep

Belugas are normally slow swimmers. They usually travel at about 1 to 5 miles per hour (1.5 to 8 km/h), but can reach speeds of 14 miles per hour (23 km/h) if pursued. They are very supple and can operate their tail so that they can swim backward, allowing them to maneuver in very shallow water that just covers their bodies. They move with a gently undulating motion that can make them look like ghostly apparitions in murky waters. They surface to breathe about every 30 to 40

seconds, but studies have shown that belugas routinely dive for periods of nine to 13 minutes. A beluga that swam up the Rhine River in 1966 was seen to be submerged for 70 minutes. Surface ice is not a great problem to belugas, since they can travel up to 1 to 2 miles (1.5 to 3 km) underwater. But they can become trapped in the sea ice, which makes them easy prey for hunters and polar bears. Belugas often carry scars from polar bear attacks.

For most of the year belugas remain offshore, near sea ice. They move closer to shore in summer, since rivers are not frozen and can be entered to find food. After mating they move to warmer waters to give birth and then migrate back to colder areas when the calves are strong enough to cope. Migrations appear to be affected by the distribution and abundance of prey and the extent of pack ice.

⊕ Belugas are among the most social of whales. Males and females usually form separate groups that join together during migrations in huge herds containing hundreds or even thousands of animals. However, the smaller family units remain distinguishable.

Some populations migrate long distances, while others live permanently in quite small areas.

It is very rare to see a solitary beluga, since they are highly social animals. The strongest social bond is between a mother and her calf: She cares for her young over several years. Although the breeding cycle is normally two years long, mothers sometimes suckle their young for up to 24 months, and extend the time between births to three years. Groups consisting of a mature female, her newborn calf, and several of her most recent young are common. There are also separate groups of about three to 15 males that will merge with female groups during the breeding season. In

Russian offshore waters the male groups sometimes contain up to 500 animals. During migrations the different groups join together in herds that can contain hundreds or even thousands of belugas. These migrating groups can extend over 6 miles (10 km) or more, but it is still possible to make out the many smaller family groups. Some adults have been seen ahead of migrating groups, appearing to explore passages through pack ice for others to follow. Larger aggregations also form in places where food is abundant.

The beluga is one of the most vocal of the toothed whales and is sometimes known as the "sea canary." It has a fantastic repertoire of trills, squeaks, moos, and twitters. What is amazing about the noise produced by belugas is that it can easily be heard above water as well as below. Underwater the sound of a herd of belugas is apparently comparable to a noisy barnyard. The sounds are used to communicate with other whales. The beluga also has a wide range of facial expressions, which are thought to be another form of communication.

Like other toothed whales, belugas produce clicks that are used for echolocation (called sonar when it takes place underwater), bouncing echoes off prey and the seabed to help the animals find their way around. The waters where belugas live are often covered with ice, and there is sometimes no sunlight for months at a time. Visibility is therefore often quite poor, and sonar is then a useful way of avoiding obstacles and hunting for prey. The bulging forehead (often called the melon) of a beluga is apparently used to help with echolocation, but exactly how is not fully understood. There is some evidence that suggests the beluga uses its sonar abilities to find breathing holes in packed ice. It is thought that they have the most effective echolocation system of all cetaceans.

Diverse Diet

The beluga's diet is diverse and includes worms, crustaceans, mollusks, and fish. Most of its food is found on the seabed at depths of up to 1,500 feet (500 m). Its mobile neck allows it to scan the ocean floor for food. Having detected prey by sight or sonar, its lips produce suction to draw the victim into its mouth. Otherwise it squirts water at small animals, dislodging them from among stones. Belugas have about 32 to 40 peglike teeth, which are not fully grown until the whale is two or three years old. The teeth often become worn and so would not be able to grasp prey. In fact, they are not used much for feeding because belugas swallow their prey whole. The teeth may wear out producing sound: The animals clap their jaws together, making loud drumming sounds that are thought to be used as a threat.

The Talkative Beluga

Belugas are very communicative animals that can produce a variety of facial expressions as well as having an impressive vocal repertoire. They often appear quite comical when they alter their lips and seem to smile, frown, or even whistle. Lobtailing, when the whale rhythmically slaps its tail on the surface of the water, is another form of communication. They will also clap their jaws together to create a drumming sound, which is thought to be a threat. Physical contact is very important to these very social animals—they love to rub against each other.

One of the predators of the beluga is the killer whale (orca). A salmon fishery in Bristol Bay, Alaska, played recorded killer whale noises underwater to keep salmon-stealing belugas away. Belugas can also become prey to polar bears and sometimes even to walruses. Polar bears attack belugas that come to breathe in small areas of open water surrounded by ice. They also attack whales that become stranded when the tide has gone out: Belugas can survive until the next tide if they have been stranded, but they are vulnerable to polar bears and people in such a situation. Humans are the main threat to belugas. Belugas use the same migration routes every year, which allows them to be exploited by hunters. The world population of belugas is thought to number about 100,000, but some populations are almost extinct because of overexploitation.

Belugas move seasonally to coastal shallow waters, where they are at risk from pollution. Another threat to belugas is that their coastal habitats are being spoiled by exploration for oil and the building of hydroelectric dams. Both activities often require detonation of explosives underwater, which has an adverse effect on the health of belugas. Ironically, the beluga will probably be saved from overhunting because industrial pollution has raised mercury levels in their meat to such high levels that they are no longer sold for human consumption.

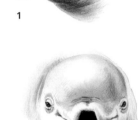

1

2

⊕ *Belugas produce loud drumming sounds by clapping their jaws together (1); the pursed mouth (2) is believed to be used when the animals are feeding from the seabed. The lips produce suction to draw prey into the mouth.*

Narwhal

Monodon monoceros

Once thought to be the horn of the legendary unicorn, the long, spiraled tusk of the narwhal makes this a distinctive and fascinating cetacean.

Common name
Narwhal

Scientific name *Monodon monoceros*

Family   Monodontidae

Order   Cetacea

Size   Length: 13–16 ft (4–5 m). Male larger than female

Weight   1,760–3,520 lb (800–1,600 kg)

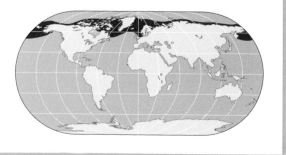

Key features
Stocky toothed whale with no dorsal fin and short flippers; skin colored with patches of gray-green, cream, and black; males have unique long, spiral tusk

Habits   Social: usually seen in groups of up to 20 animals; sometimes separate groups according to age and sex; often moves together as part of a much larger herd containing thousands of individuals

Breeding One calf born every 3 years after gestation period of 14–15 months. Weaned at 20 months; sexually mature at 6–8 years. May live 30–40 years

Voice   Clicks, squeals, and whistles used for communication or navigation

Diet   Mostly fish, squid, and shrimp

Habitat Cold arctic seas, generally near sea ice; in summer sometimes seen in estuaries, deep fjords, and bays; migrates when habitat is unfavorable

Distribution Coastal; mainly western Greenland to mideastern Canada

Status   Population: about 25,000–30,000; IUCN Data Deficient; CITES II. One of the less abundant whales, status uncertain

THE NAME NARWHAL DERIVES from the Norse word meaning "corpse whale." It could have arisen due to the animal's habit of swimming belly up and lying motionless for several minutes. Its peculiar mottled skin also makes it look like a rotting corpse. Narwhals are similar in shape to the beluga; both species lack a dorsal fin as an adaptation to life in the cold arctic seas. A dorsal fin increases the surface area of the whale and would thus speed the rate of heat loss. It could also become damaged in the ice-packed waters. Thick layers of blubber provide insulation over the rest of the body to keep narwhals warm in the icy cold waters.

Unicornlike Horn

The narwhal is renowned for having a long, spiraled tusk. From medieval times seamen and traders distributed the spectacular tusks around Europe. People believed they were unicorn tusks with magical powers that could detect if their enemies had poisoned their food. Actually, the narwhal tusk is an extremely long incisor tooth that protrudes from the left-hand side of the upper lip. The tusks are generally found only in males. A small number of males sometimes grow a second tusk on the right-hand side. Occasionally females grow a single thin tusk too. The spiral pattern is believed to ensure that the tusk will grow straight, preventing interference with swimming.

Sometimes the tusks can reach astonishing lengths. A 10-foot (3-m) tusk weighing 20 pounds (9 kg) on a 15-foot (4.5-m) whale is not unknown. There have been many ingenious ideas to explain the purpose of this extraordinary structure. Some believe it to be a weapon with which to attack other narwhals or even boats. Others suggest that it is used to make breathing holes in ice or to spear food. It is unlikely to be a weapon, since narwhals are social animals and do not tend to be aggressive. We can also rule out the other two ideas: If the tusk were needed for feeding and creating breathing holes, then females would have them too. It is now generally thought that the tusk is an ornament carried by males and used in jousts with other males to establish dominance hierarchies, like antlers in deer.

The narwhal is extremely social, forming herds that migrate when their habitat becomes unsuitable—for example, when the seas freeze over in the fall. When migrating, hundreds or even thousands of narwhals may travel together. However, these large aggregations actually consist of many smaller groups containing whales of a similar size or sexual status. Migrating in such large numbers inevitably attracts the attention of hunters. The predators of narwhals include polar bears, killer whales (orcas), and Greenland sharks; but the main threat comes from humans.

Narwhals mostly avoid waters close inshore, so are less at risk from pollutants released into the sea. However, they are still threatened by hunting. Native people have hunted narwhals for centuries. The skin, known as "muktuk," is valued because when eaten raw, it is a good source of vitamins. The unique tusks are prized, since they can be sold to tourists and collectors. The meat is fed to sled dogs, and the blubber can be used to produce oil for heating and lighting. A layer of blubber 4 to 6 inches (10 to 15 cm) thick will yield at least 100 gallons (455 l) of oil. Traditionally, the Inuit hunted narwhals from kayaks using harpoons, but some modern hunters have fast motorboats and rifles, which increase the death toll. Populations must be monitored carefully to ensure the narwhals are not overharvested.

Sperm Whale

Physeter catodon

Common name	Sperm whale
Scientific name	*Physeter catodon*
Family	Physeteridae
Order	Cetacea
Size	Length: male 49–62 ft (15–19 m); female 26–39 ft (8–12 m)
Weight	Male 51 tons (45 tonnes), maximum 65 tons (57 tonnes); female 17 tons (15 tonnes), maximum 27 tons (24 tonnes)
Key features	Largest toothed whale; dark-gray to dark-brown skin with white patches on belly; skin has a wrinkled appearance; often scarred; large, square-ended head; dorsal fin reduced to a small, triangular hump; short, paddle-shaped flippers
Habits	Females and young live in breeding schools, young males in bachelor schools, both with 20–25 individuals; older males solitary or in small groups; join breeding schools to mate
Breeding	Single calf born every 4–6 years after gestation period of 14–16 months. Weaned at 1–3 years, sometimes longer; females sexually mature at 7–13 years, males at 18–21 years. May live at least 60–70 years
Voice	Clicks used for communication and echolocation
Diet	Mostly squid; also cuttle, octopus, and fish
Habitat	Deep waters, often near the continental shelf; females and calves stay in warm waters, males migrate to colder feeding grounds
Distribution	Found in all the oceans of the world
Status	Population: estimates vary from 200,000 (minimum) to 1.5 million (maximum); IUCN Vulnerable; CITES I

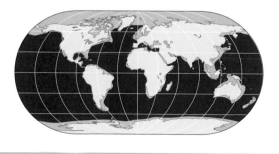

For many years the image of the great sperm whale portrayed by whalers filled people with terror. But a greater knowledge of these amazing creatures means they are now viewed with admiration.

IT IS RELATIVELY EASY TO IDENTIFY the sperm whale, since it is the largest of the toothed whales and has a distinctive, flat-ended head. The head actually contains the largest brain in the animal kingdom, although it only makes up about 0.02 percent of the animal's total body weight. The unusual shape of the head includes a forehead that forms a boxlike cavity large enough for a car to fit inside. The cavity actually contains a huge mass of spermaceti, an oily material that was once highly prized for lubricating fine machinery such as clocks. It is unclear what use the spermaceti organ is to the whale itself. Some believe it may control buoyancy when diving for food; others suggest that it may help with the animal's echolocation system.

Wrinkled Skin

Sperm whales are also unique because they have very distinctive skin that is wrinkled like a prune. The wrinkling too may have an important function, perhaps to reduce turbulence along the animal's body, assisting friction-free passage through the water. The flippers are relatively short, and the dorsal fin is reduced to a rounded hump. A series of bumps runs from the hump to the tail flukes. The powerful tail allows the sperm whale to travel at speeds up to 23 miles per hour (37 km/h). Only the lower jaw has true teeth, with about 20 to 26 pairs. They are simple, like pointed pegs, and can grow up to 8 inches (20 cm) long. Some rudimentary teeth are present in the upper jaw, but they remain

⬆ The largest of the toothed whales, the sperm whale was once a prize catch for whaling ships. Its forehead contains a mass of spermaceti, one of the most valuable of all whale products. A single whale could provide 15 barrels of spermaceti oil.

hidden in the gums and rarely erupt. However, the discovery of perfectly healthy whales that have either no teeth or deformed jaws has led scientists to believe that the teeth are not essential for feeding. It is thought that sperm whales suck prey into their mouths and swallow them whole. Their powerful jaws and teeth can be used when defending themselves, and mature males may use them when competing with rivals for sexually receptive females.

Differences between the Sexes

Male sperm whales are significantly larger than the females, a situation known as sexual dimorphism (literally meaning "two forms.") The difference in size is probably due to the fact that their mating system is polygamous, meaning one male mates with several females. Males compete for the opportunity to mate with a group of females, so they need to be big and strong to fight off rivals. The male's teeth are larger and more numerous than those of the female, possibly because they are used in conflict with other males.

The differences between the sexes are not just physical. Both sexes migrate seasonally between their feeding and breeding grounds, but the movements of females and juveniles are much less extensive than those of the males. Females and juveniles stay in warmer waters and do not travel to latitudes greater than 40° north or south of the equator. Mature males move much farther and can be found at latitudes 65° north and 70° south, close to the arctic and antarctic pack ice.

Females live in breeding groups consisting of about 12 closely related females and their young. Males leave these breeding groups when they are approximately six years old and form bachelor groups. As they get older, males tend to become increasingly solitary, only joining females briefly when it is time to mate. Both sexes feed primarily on squid, and many bear scars from battles with giant squid. Their diet also includes fish, but more so in the males—possibly because they make deeper dives than the females, where they find meaty sharks and rays near the seabed.

Sperm whales can dive to enormous depths. They are able to make such extensive dives owing to large quantities of a pigment called myoglobin in their muscles. Myoglobin stores oxygen, keeping the muscles operating when the animal is underwater and cannot breathe. During deep dives the heartbeat slows down, the lungs collapse, and the whale relies on the vast amounts of oxygen stored in its muscles to supply other vital organs. Deep

waters are dark and often murky, so the whales cannot rely on their eyes for getting around. Instead, they use sound to direct them and help them find food. When diving, sperm whales appear to be more dependent on sound than vision. Their eyes are small and inconspicuous, and their two nasal passages each have very different functions: The left is for breathing and the right for the production of sound.

Sperm whales produce clicks that are used for the echolocation of prey and as a method of communication. It is thought that their large head helps focus the clicks into a beam of sound—like a flashlight beam—to detect prey and obstacles in the underwater darkness. It has even been suggested that sperm whales use sound to stun or kill their prey by zapping it with an intense beam of sound energy. Males announce their authority with loud clicks, and a mother uses clicks to keep in contact with her calf. Scientists can use the sounds produced by sperm whales to locate them. Unfortunately, the sperm whales' reliance on sound makes them susceptible to noise pollution created by humans, including the sound of ships' engines and oil-drilling equipment.

There are strong social bonds between females and young in the breeding groups. Young calves are unable to make prolonged dives; so while a mother is diving to feed,

Olympic Divers

The sperm whale deserves a gold medal for its amazing diving skills. An adult animal can stay submerged for up to two hours at a time. It can also dive deeper than any other mammal in the world. Sperm whales dive to great depths to hunt for their favorite food, bottom-dwelling squid. Accurate sonar recordings reveal them diving to depths of 4,000 feet (1,200 m). However, analysis of the stomach contents of male sperm whales often reveals prey species typically found at even greater depths, indicating that the whales can sometimes dive twice as deep. At such depths the water is permanently dark and cold, with pressures equivalent to the weight of a bus on every square foot of the animal's body. It is generally the mature males that make the extraordinarily deep and lengthy dives for which the species is renowned.

another female in the group looks after her calf. It is possible that females do not make as deep and lengthy dives as males so that they can stay in contact with their young and return to them quickly if necessary. If young were left alone, they would be an easy target for sharks or killer whales. The females within the breeding groups take it in turn to feed, so there are always some adults present to care for the young. It is even thought that they will sometimes suckle calves that are not their own. Adult sperm whales are also very protective of one another within their social group. When they are being attacked, or if a member of their group has been injured, they will assemble together, their heads facing inward, and use

⊖ *Adult sperm whales are very protective of each other. If a member of their group is injured, they will encircle it, facing inward and use their powerful tails as a means of defense. In the past such behavior proved disastrous for the whales, which could be picked off one by one by whalers.*

spermaceti oil, one of the most valuable of all whale products. Ambergris, a substance found in their intestines, was formerly used in the perfume industry. It was extremely valuable to the whalers—a piece weighing 250 pounds (113 kg) was worth more than the wages for the entire crew for a whole year. Today synthetic substances are used. Sperm whales are now protected by international agreement.

Literary Character

In the novel *Moby Dick* by Herman Melville the sperm whale was portrayed as a monster of the sea. Hunting sperm whales was once very dangerous and was looked on as an act of great bravery. The whalers would row in an open boat to spear the whales with hand-held harpoons. The whale would sometimes turn on the whalers in self defense. It could knock them out of their small boat, a fraction of its own size, and crush them in its powerful jaws. The whalers feared the mighty tail that would be raised high above the water before the whale made a deep dive, calling it "the hand of God." However, the advancement of whaling technology, notably more powerful steam- and diesel-driven boats and the exploding harpoon gun, denied the sperm whales any chance of defending themselves. They were slaughtered for many years before being protected by international law. Because whalers selectively hunted the larger males, they created an uneven ratio of males to females. The combination of the naturally slow population growth and the selective hunting of the larger males has resulted in low calving rates. That has slowed the recovery of the sperm whale population, despite worldwide protection.

their powerful tails to defend themselves and their weakened comrade. Individuals will even put themselves at risk to try to rescue a companion that is in danger.

As the largest of the toothed whales, the sperm whale was a valuable catch to any whaler, not only for its plentiful supply of blubber, but also the provision of spermaceti and ambergris. Whalers once believed that the material in the forehead was like the sperm produced by male mammals. It is from that mistaken belief that the sperm whale and its oil take their name. A single sperm whale could provide whalers with 15 barrels of

⊕ *Whaling was once a dangerous activity. The risks faced by whalers were described in the novel* Moby Dick.

Common name Northern bottlenose whale

Scientific name *Hyperoodon ampullatus*

Family	Ziphiidae
Order	Cetacea
Size	Length: 23–30 ft (7–9 m). Male larger than female

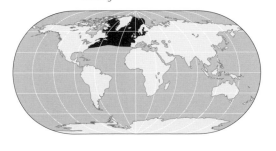

Weight	Male 8 tons (7.5 tonnes); female 6.6 tons (5.8 tonnes)
Key features	Medium-sized whale with distinctive, bulbous forehead and narrow snout ("beak"); lower jaw of beak extends slightly further than upper; 2 main teeth at tip of lower jaw in males; dorsal (back or spinal) regions dark gray to brown, lighter on belly; no notch in flukes (tail fin)
Habits	Usually lives in groups of 1–4; larger herds formed when migrating and in the breeding season; older males often travel alone
Breeding	One calf born every 2–3 years after gestation period of about a year. Weaned at about 12 months; females sexually mature at 8–12 years, males at about 7–9 years. May live approximately 30–40 years
Voice	Little known, but uses clicks and whistles
Diet	Mainly squid; also fish, sea cucumbers, cuttle, starfish, and prawns
Habitat	Cold-temperate and arctic waters, preferring offshore areas with water depths of over 3,300 ft (1,000 m)
Distribution	North Atlantic oceans and arctic regions, avoiding shallow waters
Status	Population: probably a few thousand; IUCN Lower Risk: conservation dependent; CITES II. Insufficiently known; has declined due to hunting

Northern Bottlenose Whale

Hyperoodon ampullatus

One of the longest and deepest divers of all whales and dolphins, the northern bottlenose whale has been recorded diving to depths of 4,500 feet (1,370 m).

THE NORTHERN BOTTLENOSE WHALE is one of the beaked whales. It has a long, cylindrical body and a narrow, pointed snout (or "beak") quite similar to that of the bottlenose dolphin. Its distinctive, bulging forehead is more prominent in older males and may overhang the beak. If the bulbous forehead can be seen, the whales are easy to identify, since they are the only beaked whales in the North Atlantic with such a rounded profile to the head. Adults are dark gray to brown on their dorsal (back or spinal) regions and are a lighter gray or creamy brown on the underside. Older males are easily recognized, since their forehead and beak become white, and occasionally their entire body becomes yellow-white in color. Juveniles are black to chocolate brown. The bottlenose whale has two main teeth, but some males have four, which are located on the tip of the lower jaw and are not exposed when the mouth is shut. These teeth do not usually erupt in females, but remain below the gums. Both males and females may have rows of tiny teeth along both jaws.

Deep-Sea Divers

The northern bottlenose whale is well known for its amazing diving abilities. Whalers have reported they can be submerged for up to two hours, but the typical dive time, without causing any stress to the animal, is 14 to 70 minutes. When diving, they do not travel very far horizontally and often surface again very close to where the dive began. It is amazing that air-breathing mammals can remain underwater for such a long time. They are also able to dive to extraordinary depths. Dives

⊕ *The northern bottlenose whale regularly dives to depths of 2,500 feet (760 m), and sometimes much deeper, to rummage around the seabed for its main food of squid. The whales can be submerged for up to two hours, but the more usual dive time is somewhere between 14 and 70 minutes.*

generally range from 250 to 2,500 feet (76 to 760 m) in depth, but an astounding depth of 4,500 feet (1,372 m) has been recorded.

The bulk of the diet is squid: One whale was found with more than 1,000 beaks from squid in its stomach. The whales have to use deep, sustained dives to find their food, which also includes fish and invertebrates, from the seabed. The stomachs of some individuals have been found to contain bits of shell, stones, and clay, indicating that they may use their snout to rummage around the seabed for food.

Northern bottlenose whales are widely distributed in the North Atlantic. They are most commonly found where the sea is at least 3,000 feet (914 m) deep. It is thought that their preference for deep water is due to the abundance and distribution of their prey. They are found in cold-temperate and arctic regions and prefer water temperatures that are between 32 and 62.6°F (0 to 17°C). They use the warmer waters in winter, swimming south to mate and give birth, and then migrating north again for the summer.

Loyal Friends

Northern bottlenose whales are nosy, inquisitive animals and will approach boats quite fearlessly, but mothers are extremely protective of their calves. If a calf approaches a ship, the mother will swim between it and the vessel. They are also known to be loyal animals that will stay with injured companions until they recover or die. Unfortunately, such behavior allowed them to be exploited by whalers in the past. Since they are so curious, the whalers could simply drift in an area and wait for the whales to come to them. Their numbers were drastically reduced in the 19th century when they were hunted for their blubber and the valuable spermaceti oil found in their heads. Northern bottlenose whales have been protected by the International Whaling Commission since 1977, yet there is insufficient information to make a population estimate. Although they are no longer in danger from whalers, the friendly whales are still threatened by humans through disturbance, pollution, and reduction in food owing to human activities, such as fishing.

Common name Gray whale

Scientific name *Eschrichtius robustus*

Family Eschrichtiidae

Order Cetacea

Size Length: male 39–46 ft (12–14 m); female 43–49 ft (13–15 m)

Weight 26–40 tons (22.5–35 tonnes)

Key features Robust baleen whale; fairly short, upwardly curved head; skin mottled gray, covered with patches of barnacles and whale lice; no dorsal fin, but low hump followed by series of bumps running to the large tail flukes; flippers small and paddle shaped

Habits Generally found in small groups of 1–3, but larger groups of up to 16 sometimes seen; large gatherings form at feeding and breeding grounds; performs one of the longest migrations of any mammal

Breeding Single calf born about every 2 years after gestation period of 12–13 months. Weaned at 7–8 months; sexually mature at between 5 and 11 years. May live 50–60 years, maximum documented 77 years

Voice Rumbles, groans, whistles, rasps, chirps, moans, growls, and bongs

Diet Small invertebrates scooped off the seabed, including crustaceans, mollusks, and worms

Habitat Shallow, coastal waters

Distribution Pacific Ocean; main population migrates between summer feeding grounds north of Alaska in Chukchi and Bering seas and winter breeding grounds off Baja California; smaller population found off Korea and Japan, but this group is close to extinction

Status Population: 20–25,000; IUCN Lower Risk: conservation dependent; Critically Endangered (northwestern Pacific stock); CITES I. Fairly common off western U.S.

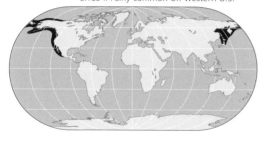

Gray Whale
Eschrichtius robustus

Gray whales make one of the longest migrations of any mammal—an annual round trip of up to 12,500 miles (20,000 km). Over one animal's lifetime the total distance covered is equivalent to swimming to the moon and back.

THE GRAY WHALE IS ONE OF the most well known and best understood of all the cetaceans, owing to its coastal habitat and friendly character. Stockier than most rorquals, but more slender than the bulky right whales, grays exhibit features that are intermediate between the two. However, they are different from other baleen whales, and as a result, they are classified in a family of their own—the Eschrichtiidae.

Barnacle Carrier
Gray whales take their name from the mottled gray color of their skin. Their bodies are also covered with patches of orange or yellow barnacles; adults may have over 350 pounds (160 kg) of barnacles attached to their skin. One type of barnacle is exclusive to the gray whale. It can be found in clusters, mainly on the head and back. The barnacles breed at the gray whales' calving grounds, so their larvae can easily find a whale to settle on.

Gray whales are one of the most parasite-heavy of all cetaceans. They are commonly infested with whale lice—small, crablike crustaceans that cling to the whale's body. Although the lice sound nasty, they can be good for the whale, since they keep wounds clean by feeding on the decaying tissue. Parasitic creatures are able to live on gray whales because their host is a slow-swimming species, and the parasites do not get washed off easily. Whale calves begin to acquire these passengers just a few days after birth.

Gray whales do not have a dorsal fin, but on the last third of the back a low hump can be found, which is followed by seven to eight smaller ones. The bowed head is small in

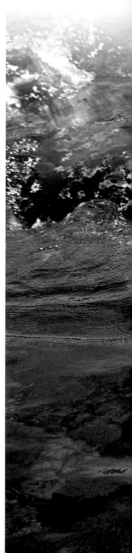

relation to the body, the tail fluke is deeply notched, and the flippers are paddle shaped with pointed tips. Grays also have much shorter baleen plates than other baleen whales because they have a different feeding technique. Although they are filter feeders, like other baleen whales, they have a rather different way of collecting their food from other species. Instead of eating floating and swimming organisms found in the open water, they feed on what they find on the seabed. They dive to the bottom in shallow waters and roll onto one side. They plow that side of their body through the sediment, then suck the mud that they have stirred up into their mouth. Using their muscular tongue like a piston, they force the sediment out through the baleen plates. They strain off any crabs, mollusks, or worms in the process and are left with a mouthful of food.

There was once a population of gray whales in the North Atlantic, but it was hunted to extinction around 300 years ago. Today gray whales are only found in the Pacific Ocean. A small population occurs off Korea, but is now Critically Endangered, having been reduced by whaling to possibly fewer than 50 individuals. The eastern North Pacific population also came close to extinction. Whalers discovered the breeding grounds in the late 1800s and slaughtered many whales. Numbers became so low that it was no longer profitable to hunt them, and they were left alone. Free from the whalers, the population began to recover, and numbers increased. However, they once again became threatened when hunting resumed with the introduction of factory ships in the early 1900s. Then in 1946 the International Whaling Commission was formed, and with gray whales listed as Endangered, commercial hunting was prohibited. They were able to reestablish themselves, and the current population is thought to be 20,000 to 25,000.

Long-Distance Journeys

Moving between the warm waters where they breed and their cold polar feeding grounds, gray whales make one of the longest migrations of all mammals. The main population of gray whales spends the summer months feeding in the cold waters of the Bering and Chukchi Seas. With the start of winter, the whales

Gray whales were once the target of whaling ships. However, today the friendly giants are a popular tourist attraction, with thousands flocking to watch their seasonal migrations along the California coast.

131

migrate south to warmer waters, where breeding and calving take place. First to leave are the pregnant females, who need to get to the warm waters to calve. They swim along the West Coast of North America until they reach their breeding grounds off Baja California. Pregnant females move into lagoons, such as Laguna Ojo de Liebre, to give birth. The warm, shallow waters of the lagoons are perfect for the newborn calves, which only have a thin layer of blubber for protection. The calves could not survive if they were born in the cold arctic seas. The lagoons are also relatively safe from the threat of killer whales, since they are shallow with narrow entrances—conditions that killer whales tend to avoid.

Friendly Whales

Grays are extremely inquisitive and friendly whales. In the lagoons of Mexico it is not unusual for a gray whale to approach a small boat of people close enough for them to reach out and touch it. They seem to enjoy having their backs scratched and will even let people reach into their mouths and stroke their tongues. Although tourism must be monitored to ensure it does not cause the whales distress or disruption, people's desire to see such magnificent creatures in their natural habitat produces support for the continued conservation of the gray whale. Their affectionate character has gained them many friends.

⊕ *Friendly and curious, gray whales will often approach small boats in the lagoons around Mexico. They will let people scratch their backs and even reach into their mouth to stroke their tongue.*

When giving birth, females sometimes have another whale with them to help. The second whale supports the mother, keeping her head above the water and sometimes will assist the calf to the surface to take its first breath. For the first few hours after birth the mother helps her calf to the surface to breathe until it can do so alone. The mother provides the calf with up to 50 gallons (227 l) of rich milk every day. The milk is 53 percent fat, and the calf will gain 60 to 70 pounds (27 to 32 kg) daily. In the warm shallow lagoons the calf establishes a strong bond with its mother, learns to coordinate its movements, and builds up its layer of blubber— all important if it is to survive the long journey to the arctic feeding grounds in the spring.

Courtship Helpers

In the breeding season males and receptive females swim together near the calving lagoons. During courtship they caress each other with their flippers. Migration is coordinated so that whales that are ready to mate arrive at the breeding grounds at the same time. It is thought that mating rituals may sometimes involve as many as five individuals that swim and roll around together. Researchers speculate that the extra whales may hold the mating pair together. Females are sexually

make the entire migration north; some may find areas where they can stay and feed for the winter without going as far as the arctic seas.

Tourist Attraction

The coastal migration of such amazing animals is now a major tourist attraction. In fact, the whale-watching industry is so lucrative that the California gray whales are now safe from hunting. Thousands of people travel to see them every year, and there are whale-watching viewpoints at strategic places on cliff tops all along the coast of California north to Oregon and beyond. The whales' high public profile offers them a secure future, but hunting is not the only danger. Noise pollution from the engines of big ships could have an adverse effect, especially during the long migration, and the dumping of sewage into the sea spoils their habitat. Also, the very same popularity that has helped save them from hunting may cause too much disturbance, especially at their calving grounds. The number of Californian gray whales is currently fairly stable, but it needs to be monitored to guarantee continued survival.

During migration and the breeding season gray whales eat very little, if at all. They live on

⊕ *When first born, gray whale calves may need their mother's help in reaching the surface to breathe. She will support the calf on her back until it can breathe unaided.*

receptive for about three weeks and probably mate with a number of different males.

When winter is over, the whales migrate north, back to the cold waters where they can find plentiful supplies of food. On the return north it is the recently impregnated females that lead the migration; last to leave the breeding grounds are the females with calves. Mothers and calves remain at the lagoons for as long as they can to allow their offspring to become as strong as possible for the long journey ahead. Swimming at less than 5 miles per hour (8 km/h), they travel along the coast, covering about 50 miles (80 km) a day. The yearly round trip of some gray whales may be 12,500 miles (20,000 km). Not all whales will

The Right-Handed Whale

Gray whales feed mainly on small organisms that live on the ocean floor. To feast on the abundant supply of food, they dive to the seabed, turn on their side, and suck up mouthfuls of their prey. It has been found that most gray whales turn on their right-hand side to feed. With that side of their head placed on the seafloor, the right-hand baleen plates become worn down faster and are sometimes shorter than those on the left. The right-hand side of the body is often scarred since it is scraped along the ocean floor when feeding, but it also means that there are fewer skin parasites on that side. Their preference for turning onto the right side when feeding is rather like the human bias toward right-handedness.

the store of blubber that lies under their skin. When the whales arrive at their summer feeding grounds, they may have lost up to one-third of their total body weight. It is during the long daylight hours of the arctic summer that the gray whales do most of their feeding. From about May to November they gorge themselves on the abundant supply of food that surrounds them. They store enough energy to allow themselves to survive for the rest of the year.

The gray whale's feeding method, plowing through the sediment to find food, may help increase the productivity of the ocean, since it releases nutrients from the seabed. The whale's specialized feeding technique allows it to exploit the seasonal abundance of food present on the ocean floor after the retreat of the arctic pack ice. Although they are primarily bottom feeders, gray whales will also sometimes eat planktonic creatures and small fish out in the open water column, as well as the invertebrates that graze on long fronds of kelp.

Grays are among the more active of the large whales and are often seen spy-hopping (lifting their head out of the water to look around), lobtailing (waving their tail fluke in the air), and breaching (leaping out of the water). They sometimes wave a flipper out of the water, almost like a greeting. Grays also produce a wide variety of vocalizations, including grunts, clicks, moans, knocking noises, and whistles. However, their sounds do not appear to be as complex or socially important as those made by other whales.

Conflicts with Killer Whales

Despite being almost twice their size, gray whales are sometimes attacked by killer whales, and many bear the scars of past encounters with these predators. Females and their calves are most vulnerable. A pack of killer whales will try to separate the pair in an attempt to get to the defenseless calf. Grays are strong animals, but killer whales are nevertheless a serious threat. Experiments have shown that grays will swim away or try to hide when they hear recordings of killer whale vocalizations.

Whales sometimes become disorientated and swim into shallow waters: When the tide goes out, unable to maneuver themselves back to deeper waters, they can become stranded on the beach. Without the water's support large whales are crushed by their own weight and cannot breathe. Being coastal, the gray whale is relatively at ease in shallow waters and often appears able to survive such strandings. It may wait quietly for the tide to come back in, when the water will refloat it. Not all are lucky enough to survive, but groups are able to withstand a few hours out of the water. It is thought they may retreat to the shallows as a defense against attacks from killer whales.

As a predominantly coastal species, the gray whale was extremely vulnerable to the whaling industry. Today their coastal habits help protect them because they are such a huge tourist draw. Many flock to see the magnificent creatures as they make the extensive migration for which they have become renowned.

⊖ *A gray whale calf breaching. During breaching a whale will lift half or more of its body out of the water, then fall back on its side into the sea.*

The Rescue of J. J. the Baby Gray

On January 11, 1997, a seven-day-old female baby gray whale was found beached on a seashore in Marina del Rey, California. It is possible that she and her mother had not bonded, and that the stranded baby then accidentally beached herself. The baby was taken to SeaWorld, where she was named J. J. in memory of Judi Jones, who had been a prominent member of "Friends of the Sea Lion." Being such a young whale, J. J. had to be fed every three or four hours to survive. All the hard work paid off when, 14 months after she was rescued from the beach, J. J. was released back into the wild. During her stay at SeaWorld she had gained about 7 tons (6.3 tonnes) and grown over 18 feet (5.5 m).

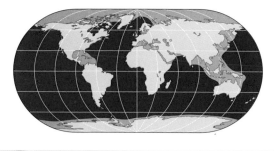

Blue Whale

Balaenoptera musculus

Common name	Blue whale
Scientific name	*Balaenoptera musculus*
Family	Balaenopteridae
Order	Cetacea

Size Length: 80–100 ft (24–30 m). Female generally larger than male

Weight 114–136 tons (100–120 tonnes), occasionally up to 216 tons (190 tonnes)

Key features Long, streamlined rorqual whale—largest animal on earth; blue-gray with pale mottling; ridge runs along top of flat, "U"-shaped head; 2 blowholes with fleshy splashguard; tapered flippers up to one-seventh of body length; small, stubby dorsal fin; tail flukes broad and triangular

Habits Shy and wary; mother and calf travel together, otherwise tends to be solitary; sometimes larger numbers found close together feeding or migrating; may associate with fin whales

Breeding Single calf born after gestation period of 10–11 months. Weaned at 7–8 months; sexually mature at 5 years in females and just under 5 years in males. May live 80–100 years

Voice Loud, low rumbling calls that travel long distances underwater

Diet Principally krill, but also other small crustaceans and fish

Habitat Mainly open ocean, but will come closer to shore to feed or breed; migrates between polar feeding grounds and warmer subtropical and tropical breeding grounds

Distribution Found in all oceans of the world

Status Population: 3,500; IUCN Endangered; CITES I

Bigger than any of the dinosaurs, the blue whale is the largest creature ever known to have lived on the earth.

WEIGHING MORE THAN 20 AFRICAN elephants, the blue whale is gigantic—the biggest animal on earth. The largest known dinosaur, *Argentinosaurus huinculensis*, was about 100 feet (30 m) long and weighed a massive 89 tons (81 tonnes), but the blue whale has been known to attain lengths of 108 feet (33 m) and weigh nearly 180 tons (163 tonnes). A heart the size of a small car pumps roughly 6 tons (5.4 tonnes) of blood around the body. The main artery, the dorsal aorta, is large enough for a human to crawl through it. The mouth of the blue whale is so big that an entire football team could stand inside on the tongue. In fact, the tongue alone weighs nearly 4 tons (3.6 tonnes), as much as a school bus.

Abundant Food Supplies

Blue whales have been able to grow to such incredible sizes by making use of the plentiful supply of food present at their fertile polar feeding grounds. Their size is possible because their body is supported by water, and so they do not require the large, heavy, and impractical bones that a land animal of equivalent size would need. Furthermore, support from the water is spread out evenly over the whole of the whale's underside, not concentrated on the bones of the hips and shoulders.

The blue whale has a slim, streamlined shape with a girth that is less than that of an adult right whale. But when it feeds the 80 to 100 throat grooves, which run from the chin to the navel, expand and increase the whale's volume as it takes in 36 to 45 tons (33 to

⬆ An underwater view of a blue whale feeding. The throat grooves can expand to increase the whale's volume, enabling it to take in immense quantities of water and krill—its main food.

41 tonnes) of food and water. Surprisingly, the largest creature in the ocean feeds on one of the smallest—krill. These tiny crustaceans are only a couple of inches (about 5 cm) long, but they are the main energy source of blues. The krill are filtered out from huge mouthfuls of water by the whale's baleen plates.

Nickname

The blue whale takes its name from the slate-blue color of its skin, which is mottled with gray and white blotches. Algae sometimes attach themselves to the stomach of the whales, giving them a yellowish tinge. The algae are actually responsible for the animal's nickname of "sulfur bottom." Blue whales have a small dorsal fin about three-quarters of the way along the back. It is tiny in relation to the rest of the body, only about 15 inches (38 cm) high. It also varies in form from triangular to sickle shaped. The tail flukes are broad—about as

wide as the wingspan of a small aircraft. The flippers are slender and tapered. A single, raised longitudinal ridge runs along the top of the broad, "U"-shaped head from the tip of the snout to the two distinct blowholes. The blowholes are surrounded by a prominent fleshy splashguard, which helps keep water out of the whale's nostrils.

As well as being the largest animal, the blue whale is also the loudest. Its calls are emitted at a very low frequency and at a volume of 188 decibels. As a comparison, a human shout is only 70 decibels, and a jet engine at full blast is barely 140 decibels. Since decibels increase by factors of 10, the whale's call is thousands of times louder than that of a human. The loudness enables the sounds to travel for many miles underwater. The calls are highly structured, with long sequences of varied sounds, like our sentences. It is thought that the "songs" are used to communicate with other whales, especially during the breeding season. Effective communication is important because blue whales are thinly spread across the world's seas, with each one having hundreds of cubic miles of ocean to itself. Meeting to mate would be a very chancy affair were it not for the ability of the whales to tell each other where they are. Because the intensely loud calls travel for thousands of miles in deep water, there is the possibility that blue whales could even communicate across whole oceans. In addition to communication the blue whale may use sound to navigate, bouncing echoes off the seabed.

Solitary

Blues seem to be fairly solitary whales. There is a strong bond between a mother and her calf, but otherwise they are found alone or in small groups of two or three individuals. Larger groups sometimes form at good feeding places. However, we need to be cautious when looking at the social behavior of these creatures because our perception of "solitary" may not be the same as that of the blue whale itself. Being such large animals, they require a great deal of space. Therefore, what we may think of as a lonely blue whale may actually, from its own perception, be in "company" with other whales that are only a few miles away. From time to time blue whales are found in association with fin whales, probably drawn together by a shared interest in abundant food.

In winter blue whales migrate from their polar feeding grounds to warmer waters to breed and calve. Little is known about mating in blue whales, since they are shy creatures and hard to locate in the open seas. Pregnancy is unusually short for such a large animal, only 10 to 11 months. When the calves are born, they are about 18 to 20 feet (5.5 to 6 m) long and weigh 1.8 to 3.6 tons (1.6 to 3.2 tonnes). The mother produces over 450 pounds (200 l) of milk every day to nourish her calf. Her milk is rich and creamy, containing 35 to 50 percent fat, and the young calf will gain more than 200 pounds (90 kg) a day. The mother and calf

stay close together, since blue whale calves are occasionally the target of a pack of killer whales. After only seven or eight months the calf is weaned, weighing at least nine times its birth weight and having doubled in length. The weaning of the calves coincides with migration back to the cold waters of the feeding grounds.

Most blues are migratory, but in some areas—such as the Pacific waters off Costa Rica (known as the "Costa Rican Dome") and off Baja California—blue whales are seen all year round. That could be because some whales do not embark on the full migration every year and stay behind. Otherwise, the population is perhaps permanently resident there. It could also be due to the seasonal overlap of populations: When the Northern Hemisphere whales migrate north to their feeding grounds, the Southern Hemisphere population replaces them at the breeding grounds. Blue whales have been observed feeding on krill off Baja California. It is unusual for the warm breeding waters to support large enough quantities of krill and this may explain why some whales

⬆ *An aerial view of a blue whale and two calves off the Pacific coast of Mexico.*

seem to remain all year round: Since food is available, they do not need to undertake the exhausting journey to the polar feeding grounds, hence saving precious energy reserves.

Despite their size, blue whales are fast swimmers: When alarmed, they can reach speeds of over 30 miles per hour (48 km/h). Their streamlined bodies allow them to move quickly through the water, faster than most ships. But they rarely leap out of the water, unlike many of their smaller relatives. They are also timid, which, combined with their swiftness, makes them hard to approach. It is also surprisingly hard to locate them, since they tend to stay submerged for long periods, only surfacing for a few minutes to breathe, then diving for up to 45 minutes. As a result, little is known about the everyday life of blues.

Favored Catch

Before the mid-1800s the blue's enormous size and speed were to its advantage. Whalers were unable to catch it, so the species was spared the massive exploitation suffered by other whales. However, the introduction of faster boats, improved whaling techniques, and the depleted stocks of more traditional catches led to the blue whale becoming the favored target. In fact, the animal's large size now provided a strong motive to hunt it. Whalers could extract about 120 barrels of oil from a single blue

Big Appetites!

The gigantic blue whale requires huge amounts of energy to sustain its large body. It gorges itself at its polar feeding grounds, taking advantage of the plentiful supply of krill. At the fertile polar waters the whales eat an estimated 3 to 3.6 tons (2.7 to 3.2 tonnes) every day, the equivalent of about 40 million krill. Some blue whale populations fast once they leave the feeding grounds because there is not such a plentiful supply of food in the warmer waters where they breed. Also, they must dedicate their time to mating. Instead of feeding, they obtain energy from their vast store of blubber (fat) that may weigh 54 tons (49 tonnes) per animal.

whale; and being so valuable, they soon made up almost 90 percent of the whaling industry's total catch. The slaughter peaked in 1931, when more than 30,000 blue whales fell victim to the whaling industry. The International Whaling Commission banned hunting of blue whales in 1966, when numbers had declined so much that the species was close to extinction. However, there are now concerns about the influence of other human activities on blue whales, such as pollution, habitat degradation, and increased levels of acoustic disturbance. Some people fear that numbers have fallen too low for a recovery to happen. It would certainly be a tragedy if this phenomenal animal were to become extinct because of human exploitation.

⊕ *Blue whales usually stay underwater for 10 to 20 minutes at a time before surfacing to take a dozen or so breaths. With each exhalation they spray a jet of water as high as 30 feet (9 m).*

Humpback Whale

Megaptera novaeangliae

Powerful, graceful, and playful, the humpback whale is the gentle giant of the oceans. As well as being admired for the production of amazing and complex songs, it is also well known as the acrobat of the seas.

Common name
Humpback whale

Scientific name *Megaptera novaeangliae*

Family   Balaenopteridae

Order   Cetacea

Size   Length: male 38–50 ft (11.5–15 m). Male generally slightly smaller than female

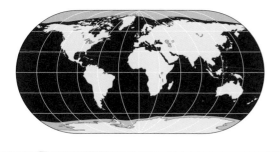

Weight About 34 tons (30 tonnes); maximum 55 tons (48 tonnes)

Key features Large, stocky baleen whale; upper body black or blue-black, underside white; long flippers; head and front edge of flippers have raised lumps called tubercles; tail flukes different in every individual

Habits   More social than other rorqual whales, rarely seen alone; congregates in large groups to feed and breed; moves individually or in small parties of 2–3 within large groups

Breeding   One calf usually produced every 2 years after gestation period of 11–12 months. Weaned at 11 months; sexually mature at 4–6 years. May live 40–50 years, occasionally over 70

Voice   Complex underwater songs consisting of grunts, groans, rasps, twitters, and moos

Diet   Seasonal feeders on krill (shrimplike crustaceans) and small fish

Habitat   Oceanic; enters shallower tropical waters in winter for breeding

Distribution Widely distributed; occurs seasonally in all oceans and from the Arctic to Antarctic

Status   Population 30,000; IUCN Vulnerable; CITES I. Uncommon and threatened

THE HUMPBACK WHALE IS SO CALLED because it raises and bends its back in preparation for a dive, accentuating the hump found in front of the dorsal fin. Its scientific name, *Megaptera*, means "giant wings" and refers to the whale's enormous flippers. In an adult whale the flippers can reach 16 feet (5 m) long, equivalent to almost a third of the total body length. Humpbacks have a more robust and less streamlined body than other rorqual whales, which narrows rapidly toward the huge tail fluke. They are also unique among rorquals in having fleshy bumps. These knobby lumps, known as tubercles, are found on top of the head, on the lower jaw, and on the front edge of their flippers. A long, coarse hair grows out of each, and it has been suggested that they could provide an improved sense of touch.

Parasite Host

Humpback whales have rough, knobby skin. In cold waters the bumps provide a home for barnacles, which cannot attach to the smoother skin of other whales. One species attaches itself deeply into the whale's skin so that only the crown shows, while another sits on top of the tubercles. When the humpbacks move to their warmer breeding grounds, the barnacles drop off, leaving scars where they were attached. The whales are now free from the barnacles, but become infested by whale lice instead.

An interesting feature of humpback whales is their tail flukes. Each whale has a uniquely shaped and colored tail fin. Similarly, the dorsal fin is different in each animal. In fact, the tail fluke and dorsal fin are as unique to each

individual whale as a fingerprint is to a human. Researchers can use these features to identify, photograph, and catalog the whales, allowing each to be individually monitored. Wherever the whales turn up, they can be recognized from the color and shape of their tail fluke. It therefore becomes possible to track a whale's movements around the ocean and provide valuable information about migration, behavior, breeding, and population sizes. Researchers often use the markings on the tail fluke to name the whales. Sometimes the name reflects dangers faced by humpbacks: "Tidbit" has a chunk of its tail missing from a killer whale attack, and "Lopsided" has lost one-half of its tail,

"Breaching" is one of the most spectacular humpback behaviors. The whales lift almost their whole bodies out of the water, twisting as they do so, and come crashing down onto their backs.

possibly as a result of injury from a ship's propeller. Ships are a threat to humpbacks, since the whales show little fear of them; they have even been known to scratch their backs on the hull of stationary vessels.

Polar Feeding Grounds

The humpback whale spends the summer in cold polar feeding grounds. It migrates to coastal tropical or subtropical breeding areas in the winter, often traveling thousands of miles. It is thought that the whales do not cross the equator, so the populations of the Northern and Southern Hemispheres are probably separate. The humpbacks of the Atlantic and Pacific do not mix either, and there are also physical differences between them: The flippers of the Atlantic humpbacks are white on both sides, sometimes with black markings. The flippers of the Pacific humpbacks are black on the upper side and white on the underside.

Humpbacks are seasonal feeders, eating huge amounts of shrimplike krill, plankton, and small fish throughout the summer when they are living in highly productive cold seas. During their stay in the warmer breeding grounds the whales do not feed, but instead spend their time mating and calving. They use energy stored as fat in the thick layers of blubber, which they have laid down over summer in the cold polar feeding grounds. Humpbacks have the most diverse and spectacular feeding techniques of all baleen whales. They sometimes hunt cooperatively, rounding up large groups of prey and gulping huge mouthfuls of water. The pleated throat grooves can expand, allowing large volumes of water to be taken into the mouth. They expel the water when the mouth closes, sieving off the small food items on the bristle-fringed baleen plates as the water is forced between them. The prey remains trapped inside the whale's mouth, ready to be swallowed. Another method of feeding used by humpbacks is called bubble netting. The whales will swim around in a spiral beneath their prey. They then start to produce bubbles by blowing air out of their blowholes. A wall of bubbles surrounds the prey and traps it, allowing the whales to swallow them all. In summer at the polar feeding grounds an average-sized humpback whale will consume 1.8 to 2.2 tons (1.6 to 1.9 tonnes) of prey over about 120 days. The humpback has between 270 and 400 baleen plates, which are dark gray and up to 26 inches (65 cm) long.

Sea Acrobats

Humpback whales produce awe-inspiring acrobatic displays. One of the most spectacular behaviors is the "breach." The whale uses its flukes to produce enough upward force to lift about two-thirds of its body right out of the water. During these amazing leaps the humpbacks sometimes twist their bodies and appear to spin out of the water. They then come crashing down on their backs with a huge splash. It is not understood why whales perform such a stunt—it could be purely for play, or it might serve as a courtship display or to loosen skin parasites. Other behaviors include lifting the head or tail out of the water, flipper-slaps, and head-slaps. Slapping the water surface produces a very loud noise like a

⊕ Humpback whales feeding in the cold waters of the Chatham Straits, southeastern Alaska.

⊕ How the humpback feeds: Plankton are engulfed in a mouthful of water (a); water is sieved through baleen plates, and food held on the plate's bristles (b).

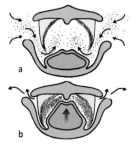

a

b

rifle shot. These loud noises are possibly used as a method of communication, enabling a scattered group of whales to keep in touch with each other, or they might act as a warning signal. The acrobatic displays are a major attraction to whale-watching tourists off Hawaii and the coast of Massachusetts in summer.

Humpbacks are slow swimmers and look extremely graceful as they move through the water. They swim individually or in small social groups that tend to be family units of about three or four. The groups keep in contact with other similar-sized groups by sound signals. Their calls can travel for hundreds of miles underwater. Large, loose groups of animals gather for breeding or feeding. However, most associations are temporary, only lasting a few hours or perhaps days before the animals move on. The exceptions are the strong bonds between a mother and her offspring, and some long-term associations between individuals living together at the feeding grounds.

by circulating air through the tubes and chambers of their respiratory system. The songs of the humpback are the longest and most varied social calls in the animal kingdom. Recordings of the magical yet eerie songs are sold all around the world. Hence, the serenading humpbacks have even reached audiences beyond their own kind.

Despite its popularity today, the humpback whale is still threatened by humans in several ways, the most obvious being whaling. It has been hunted for centuries for its oil, meat, and whalebone. It is one of the whales to have suffered most from whaling practices, since it is a slow swimmer and not easily scared. It also tends to frequent coastal waters, returning to the same regions every year. As a result, it has been easy to exploit. In recent times a ban on whaling has allowed humpback numbers to recover somewhat. Other threats to the humpback include marine pollution, depletion of food resources by fishermen, and the use of drift nets for fishing, in which the whales can become entangled.

Whale Songs

One of the most fascinating features of humpback whales is their songs. These are made up of grunts, moos, rasps, twitters, and groans that are organized in sequences. Songs are created by the sequences being repeated over and over again, sometimes for hours at a time. They can be heard up to 30 miles (about 50 km) away. The sounds are so intense and powerful that they can even be felt at close range, vibrating through the hulls of passing ships. Both males and females can produce sounds, but it is thought to be only the male who sings. Songs are generally restricted to the breeding season and so are thought to be mating calls or territorial displays. All the whales in a particular area sing the same song, but songs vary between regions. The songs also appear to change from one breeding season to the next. Whales do not have vocal cords, so it is not clear how they produce the songs. It has been suggested that they can produce sounds

Humphrey the Lost Humpback

Humpbacks migrate many miles from their cold, polar feeding grounds to reach the warmer, tropical waters where they breed. They undertake one of the longest migrations of any animal—often thousands of miles. However, traveling such long distances does not always go as planned. Humphrey was a male humpback who, in October 1985, swam more than 28 miles (46 km) beyond the Golden Gate Bridge into San Francisco Bay and up the Sacramento River. Humphrey had got lost on his way to the Hawaiian breeding grounds. Thanks to the efforts of many people, Humphrey was rescued and returned to the open sea.

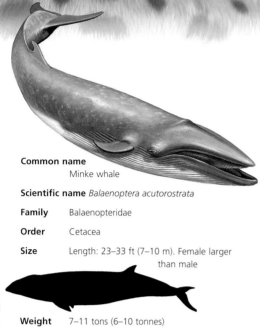

Common name
Minke whale

Scientific name *Balaenoptera acutorostrata*

Family Balaenopteridae

Order Cetacea

Size Length: 23–33 ft (7–10 m). Female larger than male

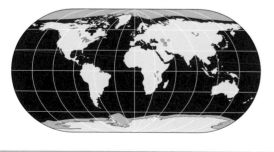

Weight 7–11 tons (6–10 tonnes)

Key features Smallest and most abundant of the rorqual whales; black or dark gray in color with white belly; top of head appears flat with a raised central ridge and distinctive pointed snout; relatively short flippers, often with white band near base; tall dorsal fin

Habits Solitary or found in groups of 2 or 3; sometimes larger groups formed when feeding; can be quite inquisitive

Breeding One calf born every 2 years after gestation period of about 10 months. Weaned at 4–5 months; sexually mature at 6–7 years. May live up to 60 years

Voice Some grunts, clicks, thumps, and pings

Diet Filter feeders; mainly krill (planktonic shrimp) and fish, but will eat squid and types of plankton other than their favorite krill

Habitat Coastal and offshore polar, temperate, and tropical waters; seems to prefer cooler regions, but sometimes migrates

Distribution Widely distributed throughout all the oceans of the world

Status Population: 500,000–1 million; IUCN Lower Risk: near threatened; CITES I

Minke Whale

Balaenoptera acutorostrata

The minke whale is the most abundant of all the rorqual whales, although estimates of its total numbers vary widely. Unfortunately, that numerical "success" has made the minke an easy target for the whaling industry.

THE MINKE WHALE HAS MANY of the physical characteristics typical of most rorqual whales, such as a large mouth with baleen plates and a flat-topped head. It is easily distinguished from the others by its small size. The rorquals are all large whales; and even though it is the smallest of the family, the minke can grow up to 33 feet (10 m) long. Its streamlined body is actually quite robust in comparison with most of its larger cousins—the general body shape is somewhat similar to that of a dolphin when viewed underwater. Relative to its body size, it has the tallest dorsal fin of all baleen whales. Minkes have slender, paddlelike flippers, which are relatively short—only one-eighth of the body length. Their skin is smooth with no barnacles, and they have a distinctive head, ending in a narrow, pointed snout. Like other baleen whales, they have a double blowhole. Minkes are very varied, and there may be three or even four subspecies.

Home Ranges

The minke is found virtually worldwide; but unlike most other baleen whales, it does not appear to undertake extensive seasonal migrations. Some minkes will move north in spring and summer and south in the fall and winter, but studies have shown that others appear to stay all year round within certain areas, known as home ranges. They are not considered a coastal species, but generally will be found within 100 miles (160 km) of land. They are also known to enter estuaries, bays, and fjords, and will move farther into polar ice fields than most other rorquals. One even swam up the Thames River through the center of

① A minke whale in a fjord in Antarctica. The northern animals belong to the subspecies acutorostrata. *The main southern form, (subspecies* bonaerensis), *often lacks a pale band on its flippers. Some scientists regard these as separate species. Another smaller southern form may be either a subspecies or species.*

London. Minkes also seem to differ from other baleen whales in that they are fairly solitary animals, being seen singly or in pairs. Larger aggregations can occur on feeding grounds.

Inquisitive Behavior

Minkes, unlike most rorquals, are inquisitive animals. They can often be seen spy-hopping: poking their heads above water to look around. They will sometimes approach and linger near ships. People on polar expeditions have been able to approach and pat minke whales that have been trapped among the pack ice. Minke whales are also quite acrobatic and can leap out of the water like a dolphin. They are relatively fast swimmers and when startled can travel up to 19 miles per hour (30 km/h). Although it was only for a few weeks, a minke whale was the first rorqual ever to be kept in captivity. An individual was kept in a netted area of sea, but soon escaped.

The minke whale is named after an infamous 18th-century Norwegian whaler. The whaler, Minke, would frequently fish for whales that were smaller than the permitted size. In time the small whales came to be known as Minke's whales. Later the name was formally given to the minke whale, since they are the smallest of the rorquals. However, small size is also the main reason why the minke whale is the most abundant of today's rorquals. For many years whalers tended to ignore them in favor of the much larger species.

The minke only became economically attractive to the whaling industry when numbers of the larger species began to decline. Threatened populations also began to receive legal protection, so whalers hunted the unprotected minke. Today minke whales are still quite common. Their population is estimated at between 500,000 and 1 million. Unfortunately, their abundance makes them vulnerable to whalers who want the whaling ban lifted to allow them to hunt minkes. Several thousand are killed anyway, supposedly for scientific study, but they still end up for sale as meat.

Mustang

Equus caballus

The mustang is the typical wild horse. With its noble bearing, graceful movements, glossy coat, and flowing mane and tail, it appears to embody the very essence of wildness and freedom.

THE MUSTANG'S APPEARANCE IS deceptive, since rather than indicating a wild ancestry, all of its enviable physical characteristics point to its domestic origins. Sadly, there are no truly wild horses left in the wild. The closest is the Mongolian wild horse, which now lives only in zoos and special reserves.

Horses have been running free in North America for hundreds of years, but they are all descended from domestic animals introduced by European settlers. Therefore they are not truly wild—the correct term for domestic animals that have reverted to nature is "feral." The word mustang comes from the Spanish term *mesteno*, which means "belonging to no one." Elsewhere in the world feral horses go by other names—in Australia, for example, members of the world's largest feral horse population (up to 200,000) are known as "brumbies." In Britain, continental Europe, and Asia there are a variety of wild ponies descended from different domestic breeds, but living free. Some are owned and are occasionally rounded up to be broken in and ridden, or they may be slaughtered for meat and hide. Others live their lives as wild animals.

Clues to Origins

Populations of feral horses often exhibit striking differences in size, build, and coat color that offer clues to their origins. For example, the mountain brumbies of southeastern Australia are robust, muscular animals, descended from escaped draft horses (called "clompers") once used in the timber industry. The American mustangs tend to bear the hallmarks of fine riding horses such as thoroughbreds, Arabians, Morgans, and quarter horses. They possess the

---

Common name
Mustang (wild or feral horse or pony, brumby)

Scientific name *Equus caballus*

Family Equidae

Order Perissodactyla

Size Length head/body: about 6.5 ft (2 m); tail length: about 39 in (100 cm); height at shoulder: about 5 ft (1.5 m)
Weight 770–1,500 lb (350–680 kg)

Key features Looks like a scruffy version of domestic horse; more graceful than true wild horses, with longer neck and legs, long, flowing tail, mane, and forelock; coat variable color

Habits Social, nonterritorial harem system; active day and night; feeds mostly by day; often seminomadic; may make seasonal migrations

Breeding Single foal (very rarely twins) born in summer after gestation period of 11 months. Weaned at 11–23 months; sexually mature at 1 year, males rarely breed before 5 years. May live 50 years in captivity, rarely over 30 in the wild

Voice Typical horse repertoire of whinnies, grunts, nickers, squeals, and whistles

Diet Grass and other plant material

Habitat Grassland, scrub, and semiarid plains

Distribution Global except Antarctica; generally restricted to marginal habitat to limit competition with grazing livestock

Status Population: 300,000 (also abundant and widespread in captivity). Protected or persecuted according to local laws

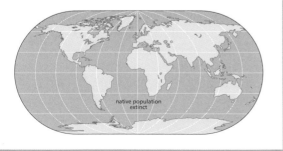

native population
extinct

⊙ Mustangs digging for water in the Hugh River, Northern Territory, Australia. Australia has the largest population of feral horses—up to 200,000.

⊙ A mustang with a domestic horse. The mustang is in fact a descendant of the domestic horse, which the Spanish brought with them to the New World.

clean legs and elegant gaits that were bred into their ancestors centuries ago by the equestrian connoisseurs of Europe and Asia. Mustangs come in all colors and include duns, bays, chestnuts, palominos, and various patterns of spots (Appaloosa) and patches (pinto). Ironically, they are never striped—that ancient characteristic of some wild horse species was lost early in the process of domestication.

Return to the Homeland

Horses originally evolved in North America and lived there continuously for millions of years before becoming extinct at the end of the last ice age. They returned in ships 10,000 years later when Spaniards occupied Mexico in the early 16th century. Within a few short years escaped animals were living wild and spreading fast throughout their ancestral home. The potential of the horse was quickly recognized by the native people, who acquired animals first by trade and theft and later by taming feral horses. By the mid-18th century there were millions of horses living wild in North America; but as settlers occupied more and more land, both horses and native people were pushed back, and both were forced to eke out a living in the poorest areas of a continent that had once been their own. By the mid-20th century free-living horses were in danger of disappearing from the American continent for the second time. They were saved by the Wild and Free-Roaming Horse and Burro Act of 1971, which protects populations not damaging the environment or harming the interests of landowners. The current population numbers 40,000 to 50,000 animals.

Back to the Old Ways

Despite their varied appearance, once returned to the wild, all feral horses revert very quickly to the lifestyle perfected by their ancestors. They are highly social animals and only rarely live alone. There are three kinds of group: The first, into which all mustangs are born, is the family group or harem. It is led by a single breeding stallion (very occasionally he will have a sidekick who helps defend the group, but does not mate any mares). The stallion collects mares as young fillies or captures them from another stallion by force. The stallion's main job is to mate with the mares, protect the herd, and lead them in time of crisis. The day-to-day movements and routines of the band are dictated by the dominant mare, usually the female that has been with the harem longest. She leads the group from their sleeping areas to food and drink, and to find salt-licks to supplement their mineral intake, and maintains discipline among subordinate mares and youngsters. Often the colts and fillies form satellite groups of one- to four-year-old juveniles that live independently, but remain close to their parents. The third kind of group is the bachelor band: young males and older stallions that have not got a harem and keep each other company instead.

Mustangs have few predators. Wolves and coyotes are too small and bears and pumas too slow to bring down a healthy adult, but they will kill an old or weak animal or foal. Feral horses are hardy, and they need to be. Those that cannot cope with life on the move or poor diet do not live long enough to breed.

⊖ *Mustangs can be any color from pale gray, like these Camargue horses, to black. Camargue horses are from the Camargue region of France.*

habitats similar to those of reindeer, where the winters can be very cold. Extreme cold threatens any bare skin with frostbite. The blood freezes, and fine crystals of ice cause tiny blood vessels to burst and the flesh to die.

Reindeer feet are broad. The cloven hooves splay out when the animal is walking, spreading its weight over a large area. Therefore, a reindeer exerts very little pressure on the ground as it walks, enabling it to travel over mud or soft snow without sinking. It is a particularly valuable adaptation for life on the open tundra, which is dominated by vast areas of soft wet moss and deep snow. The hooves also have a dished underside, with stiff hairs on part of the sole—an arrangement that gives a good grip in icy conditions.

Reindeer Moss

Reindeer feed mainly on lichens, a unique habit among deer. They specially concentrate on the pale-green tufts of so-called reindeer moss

Pestilential Insects

Biting flies and mosquitoes are a special problem in the Arctic because they become exceedingly numerous during the summer months. Vast hordes of mosquitoes attempt to bite the reindeer anywhere they can find thin skin, especially around the face. The deer are obviously hugely irritated and can be seen constantly twitching, stamping, and shaking their heads to ward off the pestilential insects. The reindeer are particularly troubled in the late summer, when the velvet covering of their growing antlers dies and begins to be stripped off, often leaving a bloody mess, which attracts the attention of biting flies. In an attempt to lessen the misery, reindeer will often seek out small patches of snow and congregate there, because the cooler air reduces the number of insects and the level of their activity.

(*Cladonia rangiferina*), which forms their principal food during the winter. The deer pull up tangles of lichens from the ground or strip them from the low branches of trees. In the spring reindeer eat a great quantity of leaves, as well as the growing shoots of low shrubs, such as dwarf willow. In the summer reindeer often consume large quantities of grass, sedges, and horsetails, particularly in the high Arctic.

During the winter, when food is buried under deep snow, reindeer often migrate south. In some places they may travel more than 600 miles (1,000 km) between the tundra, where they have spent the summer, and their main habitat at the edge of the northern forests. The large herds move at a rate of 12 to 95 miles (19 to 153 km) per day. However, individuals can run at speeds of up to 35 miles per hour (60 km/h) if necessary. The traveling herds follow traditional paths year after year. The migrations are particularly spectacular in Alaska. Not all reindeer populations are migratory. In Spitsbergen, for example, the reindeer stay in the same general area all year.

⬆ *A young reindeer resting. Unlike many other species of deer, the young reindeer's coat is not spotted.*

⬅ *In summer reindeer graze mainly on sedges, grass, and horsetails. The velvet that covers the growing antlers begins to strip off in late summer, leaving a bloody mess.*

Reindeer Calendar

Reindeer are gregarious animals. They gather into herds of many hundreds of animals during the spring in preparation for migration to the summer feeding grounds. At high latitudes summer days are long and the nights are brief. Consequently, the animals can move around continuously, feeding and resting alternately. They usually live in small, dispersed herds during the summer, each led by an old female. Males join with the herds of females toward the end of summer, ready for the rut in late September. Calves are born in the following May or early June after a gestation period lasting seven to eight months. By that time the herds are mostly out on the open tundra, where females can detect danger from a distance and more easily defend their young from predators.

Only a single calf is born each year; twins are rare. A newborn calf weighs 8.5 to 17 pounds (4 to 8 kg) and is able to walk within an hour of birth. The reindeer calf is fed by its mother on very rich milk. It contains 11 percent proteins and 20 percent fat, and helps the calf grow rapidly. The infant can be weaned at about one month of age. By their first winter young reindeer have already grown small lumps on their head, later to become antlers. Well-nourished young reindeer can also breed early, and females sometimes give birth when they are only one year old. However, the usual age at first mating is 18 months, with the first calf born when the female is two years old. Males can also become fertile in their first year of life. However, they will not breed until they are much older, since they have to compete with established bulls for access to females during the rut and are unlikely to succeed at first.

Habitat and Breeding

A calf remains with its mother for up to three years. Nevertheless, many are killed by predators, particularly wolves and wolverines, as well as eagles. Survival and breeding success owe much to the condition of the habitat. In areas of good feeding the females may breed every year. Where forage is sparse, they may have to skip years every so often in order to build up their body condition so they can produce and rear a calf successfully. Adult survival is high, and females may live for at least 15 years. As in many other deer, the male's expectation of life is less than the female's (about nine to 10 years). The male's life span is reduced by the extreme stress of the rutting period. During the rut the animals must remain highly active, yet they have very little time in which to feed themselves properly.

The reindeer's antlers are rather sprawling structures, not like the neat and compact antlers of roe and mule deer, for example. They are irregularly branched and somewhat asymmetrical. They can grow up to 4.3 feet

⊕ *In spring reindeer gather into herds of many hundreds ready for migration to the summer feeding grounds.*

Tap Dancers

As reindeer walk around, they make a highly distinctive clicking sound—like a tap dancer on a hard floor. The clicking happens even when the animals are walking over soft terrain. It is caused by small tendons in the feet being stretched tightly over knobby foot bones and suddenly released. Human joints sometimes click in a similar way, but not continuously as they do in reindeer.

(1.3 m) long, with up to 12 tines (the prongs that feature on all deer antlers) on each side.

Characteristically, the brow tines form two vertical flat plates. They are irregular in outline, like the palm of a hand, and project forward over the face in male reindeer. No other species has brow tines like them. The antlers grow during the summer. Unlike other deer, female reindeer also have antlers, but they tend to be thinner and smaller than those of the male. Males shed their antlers in December and January when the rut is over, but females may retain theirs until May. Calves begin to grow their antlers within two months of birth and

Useful Creatures

Domestication of reindeer may have begun over a thousand years ago. The arctic people of Lapland followed the herds and lived off them, much as Native American people did with herds of buffalo. Reindeer are useful creatures, since they can live in and exploit habitats that are too harsh for use by other domestic species, such as cattle and sheep. The animals provide meat, fat, milk, and cheese, as well as furs and hides from which soft leather can be made. The sinews can be turned into tough threads, and the antlers provide material for carving and making tools. Some reindeer are also used to carry loads.

In North America caribou numbers had fallen to fewer than a third of a million by 1955. However, conservation measures were introduced, and numbers steadily increased. Today the reindeer is an abundant species, with 2 million living in North America alone and at least half a million in Europe and northern Asia. There are also believed to be about 2 million semidomesticated reindeer in Scandinavia, roaming more or less freely, at least in summer.

⊕ Sleigh riding in Siberia. Reindeer can haul loads of up to 300 pounds (136 kg), a task that few other mammals could manage, especially in freezing conditions.

keep them throughout their first winter. It is not clear why reindeer alone should possess antlers in the female. It has been suggested that they might help the females defend feeding areas in winter. Scraping the snow away to uncover food supplies is grueling work, and it is vital that each animal keeps other reindeer away from its cleared space. Having antlers would enable females to clear their own snow and so to hold their own with the males. However, males lose their antlers in midwinter and are not then thought to be a threat. Indeed, they are themselves at a disadvantage during the most difficult time of the year.

Common name Elk (*wapiti*)

Scientific name *Cervus canadensis*

Family Cervidae

Order Artiodactyla

Size Length head/body: male 6.5–8 ft (2–2.5 m); female 6–7.5 ft (1.9–2.3 m); tail length: 3–7.5 in (8–19 cm); height at shoulder: 4.3–5 ft (1.3–1.5 m)

Weight Male 392–1,096 lb (178–497 kg); female 377–644 lb (171–292 kg)

Key features Coat brownish-red in summer, paler in winter; pale rump patch, dark-brown mane; antlers up to 5 ft (1.5 m) long (males only)

Habits Gregarious: lives in single-sex herds for most of year; males fight for mating rights

Breeding Single calf (twins rare) born late May or early June after gestation period of 249–262 days. Weaned at about 2.5 months; sexually mature at 28 months. May live up to 25 years in captivity, about 15 in the wild

Voice Barks, squeals; "bugle" noise (males)

Diet Grasses, forbs, bushes, and trees

Habitat Grassland, forest edge, and mountains, often near water

Distribution Western North America and parts of Asia; introduced to New Zealand

Status Population: about 1 million

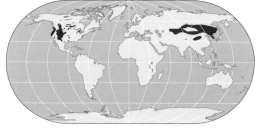

Elk

Cervus canadensis

Herds of elk once roamed all over North America, but pressure from hunting and habitat loss means they are now found mainly in the national parks and forest reserves of the West.

THE ELK IS THE LARGEST OF THE red deer group. Many people consider it to be just a subspecies of the red deer, *Cervus elaphus*, but we treat it as a separate species here. Males with a full set of 12 or more antler spikes are certainly impressive animals. Both sexes have a shaggy neck mane and a pale-buff patch on the rump that has given them their alternative name of *wapiti*, meaning "white rump" in the Native American Shawnee language.

Rescued from Extinction

Elk are found in both North America and parts of Asia. They probably reached North America around the same time as humans. Their route would have been via the Bering Sea land bridge, when sea levels were lower at the end of the last ice age. When Europeans first arrived on the continent, they found elk in huge numbers across most of the country. Estimates put the population at a possible 10 million animals. However, by the early 1900s intensive hunting was pushing the elk close to extinction. Just in time conservationists realized that the hunting was not sustainable, and now various measures are in place to secure the elk's future. Animals have been reestablished in areas where they had been exterminated. Habitats are protected and managed, particularly to minimize competition with domestic grazers, and hunting is now more tightly controlled.

There were once six different types of North American elk (subspecies or ecotypes, depending on the scientist you are talking to). Two of them, the eastern elk and the Merriman elk, are now extinct. The most widespread form is the Rocky Mountain elk (*C. c. nelsoni*), which has been introduced into sites all over North

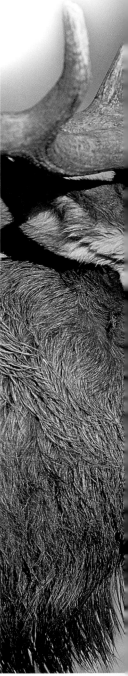

America. The Manitoba elk (*C. c. manitobensis*) tends to live on prairies and has the darkest coat. Roosevelt's elk (*C. c. roosevelti*) prefers mountainous habitat. It tends to be darker than the Rocky Mountain elk, with shorter, thicker, and heavier antlers. The Tule elk (*C. c. nannodes*) is smaller than the others.

Elk are active for 10 to 12 hours a day. Most of the time is spent looking for food, feeding, or resting and chewing the cud. Elk are large animals and need an average of 12 pounds (5.4 kg) of forage per day. They tend to feed in open areas such as meadows and grassy forest clearings. The elk's diet varies widely depending on food availability and season. In winter only tougher plants survive, and elk graze on grasses and browse on shrubs and trees such as aspen, willow, bitterbrush, and sagebrush. In summer food is more abundant, and the elk feed mainly on lush forbs such as sweet clover, lupine, and vetches. The animals store fat to make the most of the summer glut and to survive the lean winter season.

Many elk populations are migratory, traveling up to 50 miles (80 km) between summer and winter feeding grounds. Other populations are nonmigratory, but still tend to use different portions of their range in different seasons. In winter most animals move downhill as snow smothers their feeding grounds. In summer they move back up to higher ground, following the growth of lush vegetation and escaping the flies that persistently harass animals in many of the lower valleys.

Elk Herds

The size and composition of elk herds vary with season, sex, population, and habitat. The largest aggregations tend to occur in the most open areas, where in the absence of cover animals seek the security of a larger group. In summer cow-calf groups can number up to 400 animals. Cows tend to leave the group before calving. Adult males will often feed alone, if by doing so they can find good forage and avoid the aggravation of living in the herd. In August the large cow-calf groups break into smaller units and are joined by adult bulls for the rut. Each rutting group consists of up to 26 animals, with one adult bull, his harem of cows, plus calves and the occasional yearling male. Adult bulls will not tolerate the presence of bulls older than yearlings in the rutting season. At this time of year the nonbreeding males also congregate into large single-sex groups.

⊕ *A male elk in Yellowstone National Park. In the mating season, or rut, bulls impress potential mates and deter rivals with loud, high-pitched bugle calls.*

165

Wise Matriarch

Within each type of herd there is a definite dominance hierarchy. Cow-calf herds are usually led by a matriarch, an old, experienced cow who knows how to find the paths between summer and winter feeding grounds. The matriarch can also lead the herd to the good food sources. When groups are traveling, especially in winter snow, they walk in single file behind the matriarch. She communicates to them with high-pitched chirps and mews. Below the matriarch each elk has a position that is maintained by posturing and occasional fights. A high-ranking female will assert her position with ears laid flat, bared teeth, wide-open eyes, and flared nostrils. If a subordinate does not respond appropriately, she will kick out with her foreleg. Evenly matched rivals will rear up on their hind legs and "box" with their hard, sharp hooves. High status is worth fighting for, since top-ranking animals can choose the best feeding areas. They usually feed first when food is scarce. They are also able to select the best and safest resting sites.

Within the bull herds a hierarchy of dominance is also maintained. Those with the largest body and heaviest antlers are usually at

War of Strength

Battles for females are fiercely fought. Rival males stand side by side, posturing, smashing branches, and spraying urine. One charges, and the combatants lock antlers in a battle of strength that the heaviest animal usually wins. The loser has to turn quickly and run, exposing the rump (rather than the more vulnerable parts of the body) to attack. Losers are often gored, and each male may suffer 30 to 50 antler wounds per rutting season. Injuries are sometimes fatal.

the top. Most disputes are settled by posturing, but bulls of a similar size spar to establish dominance. Such tussles are more ritualized than the battles that take place during the rut. They are also much less energetic and unlikely to cause serious injury. Two rival elk will slowly approach each other with their eyes averted. They nod in unison, lower their heads, and engage antlers. With the antlers entwined, they

⊕ *A cow with calves and a bull. Adult bulls will not tolerate the presence of bulls older than yearlings in the rutting season.*

An elk drinking. Diet varies widely according to season. In summer elk are even known to feed on aquatic vegetation such as cattails.

wrestle, twisting their necks and trying to push each other around. Eventually, one animal backs off, with little outward sign of submissiveness. The gentle sparring matches help maintain peace and coordination within the herd, allowing the animals to concentrate on feeding and survival. Dominance displays go on throughout winter, well after the rutting season is over. Consequently, the bulls often keep their antlers until March. In many other deer species the antlers are only used during the rut and tend to be cast earlier.

Full-Time Courtship

The elks' mating season, or rut, starts in late August to early September. Bulls spray urine into wallows and on themselves. They also thrash their antlers on bushes and grass, and rub their neck, face, and forehead on trees to spread their scent. During the rut males spend most of their time displaying and defending their harem, and little time feeding. They rely on stored body fat, and overexertion can leave them with too few reserves to call on during harsh winters. Many die as a result. Having defended a harem, a male has to court his females. Cows choose the most impressive male to mate with and will move to another if he tires or fails to maintain their interest. Females are only receptive to males for 12 to 15 hours. If they do not mate successfully, they come into estrus again three weeks later.

When a female is due to give birth, in late May or early June, she leaves the herd to find a suitable birthing place. She will choose an area with dense vegetation that is also close to good forage. Her single calf is large, weighing about 31 pounds (14 kg) at birth. It nearly doubles its weight within two weeks as it feeds on the rich milk from its mother. A calf hides for the first 18 to 20 days of its life. Meanwhile, its mother forages some distance away, returning only for periods of nursing. The spots on a calf's coat help disguise it among the vegetation. The calf later joins a "nursery herd" and stays in a cow-calf group over the summer. Each calf has to establish its own status within the herd, and it does so mainly through boisterous play. Some elk are known to live for more than 20 years, but average life expectancy can be as low as three or four years. Males tend to die earlier, skewing sex ratios in favor of females.

Elk are nervous, vigilant animals. Their senses are finely tuned for detecting potential predators. For example, they have good vision and are particularly adept at detecting movement in the distance. In addition, the animal's hearing is excellent. If an elk hears a predator, it gives an alarm bark to alert others in the herd. With nowhere to hide in open grasslands, the elk's method of defense is to run. The major predators of elk are American black bears, pumas, and wolves. Hunting by humans is also a major factor in elk mortality.

Common name Mule deer
(black-tailed
deer)

Scientific name *Odocoileus
hemionus*

Family Cervidae

Order Artiodactyla

Size Length head/body: 5–7 ft (1.5–2 m); tail
length: 4.5–9 in (11–23 cm); height at
shoulder: male 31–42 in (80–106 cm). Female
smaller than male

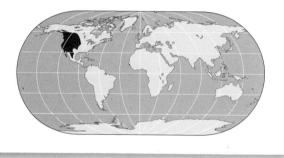

Weight 250–300 lb (113–136 kg)

Key features Medium-sized deer with large, mobile ears;
coat rusty red in summer, brownish-gray in
winter; whitish face, throat, and rump; black
bar on chin and v-shaped or triangular patch
on forehead; tail white with black tip or all-
black

Habits Active mainly at dusk and dawn; moderately
sociable: lives in small, loose herds; bouncing
gait ("stotting") when fleeing predators

Breeding Usually twins born after gestation period of
203 days. Weaned at 4 months; sexually
mature at 16 months, but males breed later.
May live up to 24 years in captivity, 10–12 in
the wild (females), 8 (males)

Voice Bleats (in fawns); snorts, grunts, and barks

Diet Grasses, sedges, forbs, bushes, and trees

Habitat Varied; usually open forest, bush, or
scrubland; often steep or rough terrain

Distribution Western North America from Alaska to
Mexico east to Nebraska and the Dakotas;
introduced to Kauai (Hawaii) and Argentina

Status Population: at least 3 million; IUCN
Endangered (Cedros Island subspecies)

Mule Deer

Odocoileus hemionus

*Mule deer are highly adaptable and able to live in a
wide range of habitats across western North America.
They have large, waggly ears, like those of mules, and
evenly branching antlers.*

MULE DEER LIVE ACROSS A WIDE swath of western
North and Central America. They can be found
from as far south as Mexico north to Canada
and Alaska and east to the Rockies. There are
11 subspecies, including black-tailed deer that
live in a narrow strip up the West Coast. Their
size and markings are variable. A large Rocky
Mountain mule deer can weigh three times as
much as an adult black-tail, but they belong to
the same species. The black chinstrap and
forehead markings and the black on the tail can
vary between populations and individuals.

Antler Growth

The bucks' antlers branch dichotomously
(evenly) to give four points per antler in adults.
Some have an additional small prong at the
front of each antler. Growth starts in April or
May, and although slow at first, by summer the
antlers can be gaining half an inch (1 cm) per
day. Growth is complete by late summer, and
the antlers are shed in winter after the rut. A
yearling male will usually have small antlers that
fork once, to give two tines. If the animal is
undernourished, the first set of antlers may only
be thin spikes. At two years old his antlers will
usually have the full four points. Afterward they
normally just increase in thickness and length.

Mule deer have a characteristic way of
dealing with predators. Rather than keeping
hidden, they use an "observe and
outmaneuver" strategy. They prefer to stay out
in the open and use their large ears and
excellent vision to spot predators in the
distance. Keeping a close eye on the enemy,
they move to safer ground, bounding
effortlessly up steep slopes where the predator
cannot easily follow.

Mule deer can reach speeds of up to 25 miles per hour (40 km/h) for short bursts and comfortably maintain a run at 18.5 miles per hour (30 km/h). When escaping, they often bounce with all four feet, behavior known as "stotting." Each bound takes them a couple of feet into the air and can carry them a distance of 10 to 16 feet (3 to 5 m). If they have to,

A mule deer buck during the rut. Large bucks mark their territory with urine and display their strength by snorting and thrashing their antlers on bushes.

they can jump fences of 8 feet (2.4 m). Their gait is ideal for rough ground, giving the animals great maneuverability. They can change direction in one bound and even reverse. They can also leap over bushes or rocks, while the pursuer must go around them.

Mule deer use a strange, stiff-legged walk when alarmed or approaching something they are uncertain about. The walk alerts others in the group to possible danger. The animals also release an alarm scent from their metatarsal glands on their legs above the midpoint of the shank.

Not So Social

Mule deer are less sociable then many other cervids. They tend to disperse widely, feeding alone or in small groups, and only come together in the rutting season or if bad weather limits their feeding ranges. Females usually stick together in small family groups, but animals within a group may not always be related. Also, the same animals may not stay together.

Groups are highly flexible, with members leaving and new members joining fairly frequently. The most mobile are the yearlings and two-year-olds. Both males and females can leave their birth group, but males are more likely to wander and tend to move farther off.

Chasing Off Rivals

The rutting season is in late fall to early winter, with the exact timing depending on the subspecies, location, and climate. During the rut bucks hardly eat at all, focusing their attention on mating and chasing off other males. Rivals are challenged to ritualized sparring matches that occasionally lead to serious fighting. A large male may keep an eye on a group of females, but he does not keep a tight enough guard over them to warrant calling the group a harem. Younger, smaller bucks may help him hold an area, hoping for the chance to mate if too many females become receptive at one time. The males wander among the females, looking for those in estrus. The females indicate their receptivity with scent glands and urine. Females may mate with several bucks over the breeding period, and a large male will usually mate all the females in his area.

Females give birth in late spring, usually to twins. For the first few weeks the fawns are dark brown and spotted. They grow quickly, and by five or six months they can weigh 66 pounds (30 kg). At one year old they weigh half as much as a fully grown adult. The fawns are fully grown by the age of eight, but males—and possibly females—may continue to grow throughout their lives.

Survival rates are low for youngsters. By the end of their first summer season 25 to 30 percent have died. By early winter only around half still survive, and only a quarter or so make it through to the following spring. Predators kill many mule deer. Golden eagles take a few youngsters, but mountain lions, bobcats, and coyotes are more of a problem. Others die of disease or accidents, perish in extreme weather, or are killed by hunters and road vehicles.

Mule deer eat a huge range of plants. They graze on grass, sedges, and forbs, and browse shrubs or trees. The nutritional quality of different plants varies with the species, the time of year, and also the type of soil the plants are growing in. Twigs are less nutritious than forbs, but at certain times of year there is nothing else. Mule deer compensate for such changes by eating as much as they can in spring and storing it as fat. They are good at picking the most nutritious plants and often supplement their mineral intake at salt licks.

Telling the Difference—Mule Deer and White-Tailed Deer

Mule and white-tailed deer (*O. virginianus*) are closely related, but have a few key differences. Mule deer antlers branch evenly, while the white-tails have small side branches off a main stem. When alarmed, mule deer prefer to remain in the open and use high, four-legged bounds over obstacles. In contrast, white-tails leap gracefully or take a leisurely walk into undergrowth to hide, lifting and waggling their tails. Mule deer hold their black-tipped tails low. Another difference is in their distribution: white-tailed deer are found mainly in the eastern United States, mule deer in the west.

A juvenile white-tailed deer buck signals deference to a dominant buck by licking its head. The prominent antlers indicate their relative social status.

A female mule deer nurses her young. Litters usually consist of one set of twins.

Habitat Preferences

Mule deer can live in many types of habitat. Typically, they are found in semiarid open forest or shrubby areas. However, they can live practically anywhere with enough different types of plants to feed on through the seasons and enough cover or high ground to escape from predators. They dislike open, exposed land, so on prairies they tend to cluster around "breaks," such as the shrubs lining rivers.

Mule deer are creatures of habit, usually sticking to relatively small, familiar home ranges. If they migrate seasonally, they tend to return to the same spot year after year. Knowing the location of every rock, bush, ledge, and steep drop is invaluable when being chased. As snow covers their feeding areas, they head down to lower ground or to exposed slopes where the snow tends not to lie. Such migrations can vary from a short walk to 100 miles (160 km). During winter, as fewer and fewer suitable feeding grounds remain, animals are forced to gather together in larger herds to share whatever food is available.

Mule deer populations have had a checkered history. There were probably about 5 million mule deer in the Americas before European colonization. Numbers took a nosedive during decades of uncontrolled hunting, from the mid-1800s to their lowest at the beginning of the 20th century. They then recovered, but from the 1950s the deer started to damage crops and plantations. Many areas were also becoming overgrazed, leading to mass starvation of the deer. Mysteriously, in the 1960s and 1970s populations went into another sharp decline. Now, careful research and population and habitat management mean that mule and black-tailed deer are no longer threatened across most of their range.

A mule deer in Glacier National Park, Montana. In winter, populations that live on mountainsides usually have seasonal migrations to lower slopes where the snow does not lie.

171

Common name American bison (buffalo)

Scientific name *Bison bison*

Family	Bovidae
Order	Artiodactyla
Size	Length head/body: male 10–12 ft (3–3.8 m); female 7–10 ft (2.1–3.2 m); tail length: 17–35 in (43–90 cm); height at shoulder: up to 6.2 ft (1.9 m)

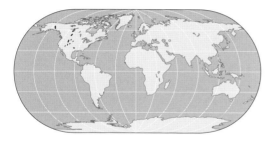

Weight Male 1,000–2,000 lb (454–907 kg); female 790–1,200 lb (358–544 kg)

Key features Large, oxlike animal with head held low and large hump over the shoulders; forelegs, neck, and shoulders covered in long, dark-brown hair; horns present in both sexes

Habits Lives in large herds that migrate across open grasslands; feeds mostly early and late in day

Breeding Single calf born May–August after gestation period of 9–10 months. Weaned at about 6 months; sexually mature at 2–3 years. May live up to 40 years in captivity, up to 25 in the wild

Voice Snorts, grunts, and cowlike noises; bulls bellow and roar during the rut

Diet Mostly grass; also sedges, wild flowers, and shrubs such as willow, birch, and sagebrush; lichens and mosses in winter

Habitat Prairies, sagebrush, and open wooded areas

Distribution Midwestern U.S. and Canada

Status Population: 200,000–500,000; IUCN Lower Risk: conservation dependent; Endangered (subspecies *B. b. athabascae*); CITES II

American Bison

Bison bison

The fate of the American bison demonstrates one of the worst examples of ruthless exploitation, which reduced the animal to the point of extinction. Conversely, it is also one of the best examples of successful conservation management.

THE BISON IS THE BIGGEST ANIMAL to have roamed the North American continent in historic times. Scientists know it by this name, but to most Americans it is more familiar as the buffalo. The two names apply to the same animal, and confusion arises as an accident of history. A similar problem applies to the use of the word "Indian" in reference to the native people of North America. Logically, Indians come from India, not America, but European explorers meeting these people for the first time called them Indians to differentiate them from white men like themselves. Similarly, when the bison was first discovered by European explorers, they sometimes called it "buffalo" because it reminded them of the water buffalo—a species that was domesticated in Asia hundreds of years ago. It was also rather similar to the African buffalo. They were the nearest familiar creatures to the newfound bison of North America. The terms "Indian" and "buffalo" have continued to be widely used to this day.

Dangers from Humans

Huge herds of bison used to roam the open plains and lightly wooded areas of central North America. It is claimed that the total population may have numbered up to 50 million animals, but they were slaughtered mercilessly by the spreading human population. As a result of such actions, the bison had already become extinct east of the Mississippi River by the early 19th century. As ranching and settlement steadily expanded westward in the United States, the bison's decline may also have been hastened by diseases caught from domestic cattle, to which the wild species was not resistant. In the Midwest commercial hunting for hides and meat resulted in massive slaughter. The coming of the railroads not only created a market for meat, but also made it possible to export meat, hides, and bones to distant buyers, increasing the pressure on the herds.

Wasteful Executions

Professional hunters like "Buffalo Bill" Cody were engaged to supply railway workers with food, and many of them killed several thousand buffalo each year. The hunting was exceedingly wasteful, since often only the skins were taken. Sometimes just the tongues were collected, the rest of the meat being left on the prairie to rot, since it was not valuable enough to warrant the cost of transporting it elsewhere. Some animals were shot from moving trains for target practice and never used at all. For every bison skin that actually reached the market, at least three other animals were often wasted. An English traveler in 1873 counted 67 carcasses at one spot where hunters had shot buffalo coming to drink along the Arkansas River. An army colonel counted 112 bodies within a 200-yard (183-m) radius, all shot by one man from the same spot in 45 minutes.

⊖ *The American bison is an unpredictable animal. It can sometimes be approached closely, but at other times will stampede at the least provocation.*

Railroad Casualties

The transcontinental railroad also split the bison herd in two, making two smaller populations and also making it easier to gain access to the animals. As late as 1870 there were still 4 to 5 million buffalo to the north and plenty more to the south. Today it is difficult to believe that so many large animals ever roamed the American plains, and people assume stories told about "millions of buffalo" must surely be exaggerations. However, there is documentary evidence proving that there were indeed enormous numbers killed. For example, fur company records show more than 35,000 bison skins being shipped from Fort Benton in 1857 alone. The Santa Fe Railroad carried over 1.3 million hides in just three years (1872 to 1874). In the north the manager of the Northern Pacific Railroad reported that his company had transported 30,000 to 40,000 skins each year in the late 1870s from Bismarck, North Dakota. In 1881 the quantities reached over 75,000. But within 10 years the trade had virtually collapsed, reflecting the almost total extermination of the bison.

Bison Census

A census in 1887 found only 541 bison left on the prairies. Conservation efforts, led by W. T. Hornaday, established captive herds in Montana and Oklahoma, and the bison has not looked back since. Today bison roam widely on the American prairies and in the sagebrush country of Wyoming. There are also large herds in South Dakota and on many private ranches. The only place where a wild population has always remained is Yellowstone National Park. About 1,500 animals live there, but sometimes range outside the park, where they damage crops and perhaps also spread disease to cattle.

European Bison

The European bison (Bison bonasus) is Europe's largest land animal and looks like its American cousin, but is taller and more slender in appearance. It is bigger than an ox, with a short, thick, hairy neck. The humped shoulders are less pronounced than in the American species, and the head is held higher. European bison are forest-edge inhabitants, coming out to graze in the open, where they eat about 65 pounds (30 kg) of grass per day. They also eat leaves and bark. In winter they are often given additional food to help their survival. The bison live in small herds of up to 20 animals. Their rutting season extends from August until October, and a single calf is born between May and July. Calves are fed by their mother for up to a year, and they can live to be more than 25 years old.

The species used to occur widely in the forests of Europe, but was brought to the brink of extinction by habitat loss and excessive hunting. By the 19th century only two populations remained, one in Poland and the other in the Caucasus Mountains of southeastern Europe. Both were wiped out early in the 20th century as a result of poaching. About 50 bison remained in various parks and zoos. Enough animals were bred from them to support reintroductions to the wild. The total number of European bison now exceeds 3,000, distributed among more than 20 wild populations and over 200 parks and zoos.

⊕ *A herd of bison graze in Yellowstone National Park—the only place where wild herds have lived continually.*

Protected from hunters and predators, bison numbers have steadily increased, and by 1995 the total population was about 150,000— almost 90 percent of them on privately owned ranches. The herds now need to be culled annually to avoid the animals becoming too numerous for their food supply to support. Bison meat has high market value, being tasty and low in fat, and many cattle ranchers keep bison as a commercial venture.

Slow Grazers

Bison are essentially grazing animals, living in large herds on the short-grass prairies, but also in lightly wooded areas. They are active during the day and also at night. They generally spend their time moving slowly, grazing as they go. They cover a mile or two (about 3 km) each day. In the past they would migrate long distances to fresh feeding areas, but that is rarely possible now, since almost all the modern herds live within enclosed areas.

Nevertheless, where there is room to do so (in Yellowstone National Park, for example), the bison still move seasonally from the high ground where they spend the summers to richer pastures on lower ground in the fall. Bison normally spend much of their time resting, but they also like to wallow in dust or mud and rub themselves against fence posts, boulders, and trees. They have acute senses of hearing and smell. Despite their large size and rather ungainly appearance, they can run at speeds of nearly 40 miles per hour (64 km/h)— at least for short distances. They are also capable of swimming across large rivers. Bison herds are normally composed of a few dozen animals, although in the past many thousands might occur in the same area. Mature males travel alone or in small groups for most of the year and join with the females for the summer breeding season.

During the rut, in July and August, dominant males fight fiercely by butting each other head to head. They make a lot of noise at that time of year, bellowing and roaring to establish status—sounds that can sometimes be heard miles away. Successful males stay close to receptive females for several days until they are able to mate with them, meanwhile keeping rival males away. A single calf is born after about 10 months—twins are very rare. The young animals become capable of breeding

⊕ Two European bison bulls sparring. The rutting season in European bison is from August to October, with one calf born between May and July. These bison are Europe's largest land animals—taller and more slender than their American cousins.

from the age of about two years, but there seems to be a geographical variation in breeding success. In Oklahoma about two-thirds of the adult females may be found breeding each year, but more than three-quarters of females do so in Montana. Females can produce a calf every year but sometimes miss a year, allowing time to build up their body reserves before becoming pregnant again.

Little to Fear

Newborn calves weigh about 35 to 70 pounds (16 to 32 kg). They can run after three hours and are weaned by the time they are one year old. The mother guards her calf jealously and will chase away predators and other intruders. Wild bison have little to fear these days now that wolves and other large predators are scarce. They are the biggest land mammals in the Western Hemisphere, and many will live to be 20 years old unless they are culled by herd managers or licensed hunters.

A smaller type of bison known as the wood bison (*Bison bison athabascae*) occurs in wooded areas of southwestern Canada. It is often treated as though it were a different species, and it has been listed as Endangered by the United States government. However, DNA (genetic molecular structure) analysis suggests that the wood bison is in fact not a separate species, merely a smaller northern race.

⊖ A herd of bison stampeding across the prairies is an awe-inspiring sight. The bison were once found in vast numbers, but hunting brought them to the brink of extinction. Now they are no longer threatened.

A Keystone Species

The bison was once the dominant factor in the ecology of the North American continent. Its grazing helped maintain short-grass prairies in a condition that was suitable for many plains species of birds, reptiles, and plants that were unable to thrive where the grass grew taller. Bison are among the natural prey of cougars and wolves. The remains of their carcasses fed scavengers, and their molted fur was eagerly collected by nesting birds. Some Native American people depended heavily on the bison herds for meat, hides, and many other useful products. The skins were used to make weatherproof tents and clothing sewn with lengths of bison

sinew. Hair was used for bedding, and bones were carved into ornaments and tools. The bison supported a whole community of plants and animals within which it lived. Removing these vital creatures from the scene disrupts the whole ecosystem, just as removing the keystone from the center of the arch of a bridge will cause it to collapse. For a while it was even official policy to remove bison in an effort to undermine Native American communities during the westward colonization of North America.

Two elderly Native American women photographed in the 1950s clad in buffalo-skin capes. Some Native Americans were heavily dependent on the buffalo.

Common name American bighorn sheep

Scientific name *Ovis canadensis*

Family	Bovidae
Order	Artiodactyla
Size	Length head/body: male 5.5–6.2 ft (1.7–1.9 m); female 4.9–5.2 ft (1.5–1.6 m); tail length: 3–5 in (7–12 cm); height at shoulder: 27.5–43 in (70–110 cm)

Weight Male 126–310 lb (57–140 kg); female 125–175 lb (57–80 kg)

Key features Brown body; white muzzle, underparts, and rump patch; brown horns—large and curled in rams, smaller and straighter in ewes

Habits Active by day; sociable: congregates in same-sex groups of 5–15 animals

Breeding Usually single lamb born after gestation period of about 175 days. Weaned around 4–5 months; females sexually mature at 4–5 years, males at 6–7 years. May live 24 years in captivity, 12 in the wild. Females live longer than males

Voice Bleating in lambs; short, deep "baa" in adults

Diet Mainly grasses; also forbs and some shrubs

Habitat Semiopen rocky terrain; alpine to dry desert

Distribution Southwestern Canada to western U.S. and northern Mexico

Status Population: 65,000–68,000; IUCN Lower Risk: conservation dependent; CITES II. Hunting now controlled, but poaching for horns continues in some areas

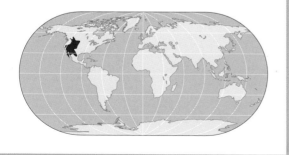

American Bighorn Sheep

Ovis canadensis

American bighorns are stocky sheep that live in the remote deserts and mountains of North America. The rams fight for dominance using their huge horns.

BIGHORN SHEEP ARE PERFECTLY built for life in extreme habitats. They live in remote, rocky places with low vegetation and few trees. There are seven geographical races (sometimes regarded as subspecies). Desert bighorns live in the dry lowland deserts of the southwestern United States and Mexico. Others live in the mountain ranges that stretch from Canada down the western side of North America.

Suited to Harsh Habitats

Bighorns have excellent eyesight and are amazingly agile considering their relative bulk. They can leap up steep, rocky slopes and walk along ledges only a couple of inches wide. Their hooves are well adapted for scrambling over rocky terrain. Their toes are independently movable, separating to grasp either side of stones. The back half of each foot has a round, rubbery pad for extra grip. Most of the plants in the animal's harsh habitats are tough, so the sheep have long, broad molar teeth for grinding and a long rumen for digesting the food as efficiently as possible. The coat has a thick fleecy underlayer. It is brown with a pale rump patch that extends down the back of the legs. The muzzle is also pale. The contrasting markings may help emphasize particular postures so that the sheep are better able to interpret the mood of others in the group.

Rams have massive horns that grow in a tight, sweeping spiral. A ram's horns are used as weapons and shields in fights, and to impress females and rival males. Females also have horns, which are shorter and straighter than those of the male. The horns start to grow when lambs are about two months old and

In fully grown males the horns can be up to 18 inches (46 cm) across and weigh as much as 13 percent of the animal's total body weight.

continue to grow for the rest of their lives, getting wider at the base, as well as longer. Displaying a set of huge horns is a sign that a ram is fit and capable of finding plenty of high-quality food—traits that a female would do well to pass on to her young.

Horns grow faster in summer than in winter, creating annual rings that can be used to assess age, at least for younger animals. Length is a less reliable age guide, since the tips are often worn down from fighting, accidents, and digging. Animals of both sexes use their horns to help clear snow from feeding areas and for grubbing up plants.

Bighorns are sociable animals, living in small groups called bands. Sometimes bands come together to form herds. Most of the time males and females live in separate bands, only linking up in the mating season. Female bands are usually five to 15 animals. Their numbers increase during the spring lambing season. Males live in bands of two to 12, with an average of five animals.

Bighorns are not territorial, but occupy home ranges whose location and size depend on group size, habitat quality, and season. They forage within the home range, wandering steadily but not always following regular trails. They normally walk about 0.2 to 0.5 miles (0.4 to 0.8 km) daily and can travel up to 2 miles (3.2 km) in a day.

Avoiding the Snow

Most populations are migratory to some extent. Some groups merely travel a mile or so up and down the mountainsides, heading to lower ground in the fall as bad weather closes in and snow covers the feeding grounds. In the spring they follow the retreating snow and fresh vegetation to higher ground. Other groups can travel over 35 miles (56 km) in a season, depending on the weather and the quality of food available. Lambs learn the long-distance routes as they follow their mothers.

Bighorns are active during the day, feeding in bouts of one or two hours before resting and chewing the cud. Daytime rest spots are usually in shallow scrapes near the feeding areas. The animals tend to feed close to "escape grounds," areas with rocky precipices and narrow ledges where few predators can follow. At night the sheep retreat to permanent bedding areas on higher ground or to caves where available—again choosing relatively inaccessible places.

The mating season occurs once a year and is called the rut. In northern regions the period is in early winter, from November to December. In the southern desert the rut can last nine months, peaking in August and September. Rams, especially the young ones, are always on the lookout for opportunities for sex. They will attempt to mount females even if they are not in estrus. If there are no females around, males will often mount each other.

Males can tell when a female is in estrus by sniffing her rear and tasting her urine. When a female is in estrus, she becomes more aggressive and determined as she searches for the male with the largest horns. Mating is usually preceded by a chase. It can be a playful, token affair or long and drawn-out, with both animals pausing to rest periodically.

Mating and Habitat

Mating patterns depend partly on the type of habitat the animals live in. On steep, rocky areas with narrow ledges males tend to stick to one female, since visibility is limited, and there is not much room for maneuvering. On more open ground a male can keep an eye on several females and will maintain a harem by chasing away other males. But a female does not always remain with one male and will search for another if "her" male is exhausted from keeping together his harem.

After a gestation period of about 175 days a single lamb is born; twins are rare. In northern parts of their range births of bighorn sheep peak in late April to late June. Before giving birth, the females move to high ground where precipitous slopes provide protection from most predators. A lamb can walk almost immediately after birth and within a day or two is as nimble footed as its mother. Within a couple of weeks a lamb will nibble at grass as well as drinking its mother's milk. Lambs are usually fully weaned by four or five months. Young males tend to leave their mother's band within one to four years. They spend some time wandering alone before

joining a band of rams. Females usually stay with their mother's band for much longer and possibly for life. They reach sexual maturity at four or five years. Rams take longer to reach their adult size, normally six to seven years.

Mortality is high in the first two years of life. Predators such as wolves, coyotes, and pumas are a serious threat. Golden eagles take a few young lambs, but are not a major predator. Other causes of death include accidents such as falls and avalanches. Such life-threatening incidents can happen to an animal of any age, but weak individuals and inexperienced young suffer most. Parasites and diseases, such as mange, lungworm, and pneumonia, also take their toll.

If the youngsters survive their first couple of years, they stand a fairly good chance of reaching "old age." The average life span is nine years, but females can reach over 20. Males tend to die earlier than females, partly due to the stresses of fighting during the rut.

Bighorns are suffering from many pressures associated with humans. Although they tend to live in remote places, disturbance from tourists can be harmful. A more serious problem is competition for grazing, both from domestic stock and feral horses, donkeys, and other grazers. Competition for resources is a particular problem in winter, when the animals tend to congregate on lower and flatter feeding areas. It is less of a dilemma in summer when the grass grows more vigorously. In the summer months the sheep can also

⬆ *Two male bighorns fight for breeding rights. The larger the horns, the more attractive a male is to a female.*

⬅ *A mother suckles her young on the precipitous slopes of the Rocky Mountains. The terrain provides protection from most predators.*

retreat to higher and rockier places that are less frequented by other species.

Present Danger

Bighorns have been hunted for many centuries. However, although hunting has been banned or controlled since the early 1900s, the animals are still poached in some areas. The dominant males are especially vulnerable, since their large horns make impressive trophies. Populations have taken a long time to recover from overhunting, partly because females have only one offspring per year.

Battle of the Horns

Males show off their horns in spectacular displays of strength. Such exhibitions are usually enough to signal status and chase off a lower-ranking individual. However, if two rivals are well matched, they will fight for dominance. They stand face to face, then run toward each other. Just before they meet, they rise on their hind legs and throw their full body weight into a crashing head butt. Their skulls are structured to absorb such huge forces, but even so, fights sometimes cause serious injuries and even death.

Common name
Muskox

Scientific name
Ovibos moschatus

Family Bovidae

Order Artiodactyla

Size Length head/body: male 7–9 ft (2.1–2.7 m); female 6–8 ft (1.9–2.4 m); tail length: 3–5 in (7–12 cm); height at shoulder: 47–59 in (120–150 cm)

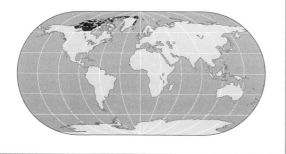

Weight Male 410–900 lb (186–408 kg); female 353–420 lb (160–190 kg)

Key features Stocky ox with short legs and neck; slight hump at shoulders; large, rounded hooves; coat black with light saddle and front; fur dense and long; sharp, curved horns in both sexes

Habits Normally active by day; also after dark on long winter nights; social: often forms herds

Breeding Single calf (twins rare) born late April–mid-June every 2 years after gestation period of 8–9 months. Weaned at 9–12 months; females sexually mature at 2 years, males at 5 years. May live at least 24 years in the wild, probably similar in captivity, but rarey kept

Voice Bulls roar, calves bleat

Diet Grasses, lichens, sedges, herbs, and shrubs such as willow and dwarf birch

Habitat Arctic tundra near glaciers

Distribution Greenland, northern Canada; reintroduced to Alaska, introduced to Russia, Norway, and Sweden

Status Population: 66,000–85,000. Now recovering after populations severely reduced by hunting in 19th century

Muskox

Ovibos moschatus

Muskoxen eke out a living in the extremely cold environment of the arctic tundra. Their dense, soft fur keeps them warm. Herds bunch together and face potential predators with a barrier of horns.

MUSKOXEN ARE STOCKY ANIMALS. The thick, shaggy coat and shoulder hump give an illusion of great size, but they are much shorter than an average adult human. Both males and females have horns, but in males the base of the horns (the boss) spreads across the whole forehead. In females it is smaller and divided by a central line of hair. The scientific name *Ovibos* (literally "sheep-ox") refers to the characteristics that the animals share with sheep and cattle. The common name comes from the smell of urine sprayed on their abdominal fur.

Scraping a Living

Muskoxen manage to survive in the arctic tundra—one of the coldest, least productive parts of the world. Here brief, cool summers alternate with freezing winters that last for eight to 10 months of the year. Much of the tundra habitat is bare, rocky ground, and the little vegetation that does exist is low and scrubby. Only the toughest grasses, sedges, and bushes survive the harsh conditions. Plants grow only slowly in such an environment, limiting the numbers of larger herbivores like the muskox that can live here. The oxen often have to scrape away deep or encrusted snow with their front feet to expose the meager plants. In the brief summer grassy river valleys support a few herbs such as alpine lettuce.

Muskoxen do not hibernate as other animals would do in such harsh conditions. Instead, they conserve energy by moving slowly and deliberately across the inhospitable terrain. Daily travel to find food is kept to a minimum, usually between 1 and 6 miles (1.6 and 10 km). Even seasonal migrations are relatively short, generally less than about 30 miles (50 km).

Body design also helps the muskox retain as much heat as possible. The stocky build minimizes heat loss and the long, shaggy coat almost reaches the ground in winter. The guard hairs, which are sometimes over 24 inches (60 cm) long, cover a dense undercoat of soft, light hair. Muskox wool is among the finest found in any large mammal. The fur covers the ears, tail, and scrotum and udder, so that no extremities are left exposed. Muskoxen also bunch together in snow storms and high winds for warmth. Although they are well adapted to the cold, starvation during severe winters—in which snow or ice covers all vegetation—is a major cause of death among the animals.

A pair of muskoxen on Devon Island, Northwest Territories, Canada. Much of the muskoxen's habitat is bare rocky ground that supports little vegetation.

Muskoxen are basically social animals. Some adult males are solitary during the summer, but most live in bull groups of two to five animals. Females and their young live in mixed-sex summer herds of about 10 individuals. In winter larger herds of up to 50 animals are formed as males join the females, and the small herds aggregate. The rutting season is from August to September, with dominant males keeping other males away from their harems using displays, loud roars, and scent marking.

Clash of the Titans

Clashes between competing males are spectacular. Facing each other, they back away, swinging their heavily horned heads from side to side. When they are far enough apart, they charge at up to 30 miles per hour (50 km/h), meeting with a head-to-head clash of huge force. The broad base of horn at the crown of the head acts as a crash helmet, providing some protection from serious injury. The clashes may be repeated for nearly an hour until one of the pair eventually backs down.

When approached by a wolf, brown bear, or other threat, muskoxen will cluster in a circle or crescent. Young animals are protected in the center, and the enemy is faced by a wall of large feet, heavy heads, and sharp horns. Their behavior works against most predators. However, muskoxen were hunted to near extinction at the end of the 19th century by a combination of settlers, professional hunters, and native peoples using more widely available firearms. Native animals in Canada and Greenland are now protected, and populations were reintroduced to Alaska. Small numbers have also been established on the high ground between Sweden and Norway.

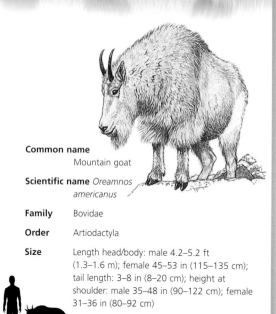

Common name
Mountain goat

Scientific name *Oreamnos americanus*

Family Bovidae

Order Artiodactyla

Size Length head/body: male 4.2–5.2 ft (1.3–1.6 m); female 45–53 in (115–135 cm); tail length: 3–8 in (8–20 cm); height at shoulder: male 35–48 in (90–122 cm); female 31–36 in (80–92 cm)

Weight Male 101–309 lb (46–140 kg); female 101–126 lb (46–57 kg)

Key features Coat yellowish-white with thick, woolly underfur; long guard hairs form stiff mane on neck and rump; horns black and curved, thicker in males; short, strong legs; short tail

Habits Diurnal, but rests during warmest part of day; solitary or lives in small groups for most of year; males fight for dominance

Breeding Single young or twins born May–June after gestation period of about 180 days. Weaned at 3 months; sexually mature at 18 months. May live about 19 years in captivity, 14 in the wild (males); females a few more

Voice Various sheeplike bleating sounds

Diet Variety of trees, shrubs, grasses, and herbs

Habitat Steep cliffs, rocks, and edges of glaciers

Distribution Southeastern Alaska and south Yukon to Oregon, Idaho, and Montana; introduced to some other mountainous areas of North America

Status Population: probably about 50,000–100,000. Not threatened

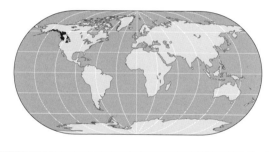

Mountain Goat

Oreamnos americanus

Mountain goats thrive in cold, craggy terrain, where their white coats make them practically invisible against the snow. Even in summer they look like small patches of snow among the dark rocks, high on the mountainous crags they inhabit.

MOUNTAIN GOATS ARE NOT TRUE goats, but are members of the group known as the Rupicaprini, which literally means "rock goat." Their nearest relatives are the goral (*Nemorhaedus goral*), takin (*Budorcas taxicolor*), and serow (*Capricornis sumatraensis*), which live in Asia, and the European chamois (*Rupicapra rupicapra*). Mountain goats probably reached America via the Bering land bridge when sea levels were lower during the Pleistocene epoch.

Functional Coat

Mountain goats are white or more often a dirty pale yellow, with black, curved horns and short, strong legs. The stockiness is exaggerated by the thickness of their hair, which makes them look squat and thickset. Yet underneath their long coat they are quite slim. The long, stiff guard hairs on the back and shoulders give the goats a pronounced "hump," and thick hair on their upper legs makes them look as if they are wearing short trousers.

However, the long hair is a vital necessity—mountain goats live in some of the coldest, least hospitable places in the United States. They inhabit mountains—usually above the tree line on the edges of major glaciers or snowfields. Their dense coats keep the animals warm in even the most biting wind. The underfur is thick, woolly, and as soft as cashmere. The long guard hairs give some protection against snow and rain, and so prevent the warm underfur from becoming waterlogged. They use their strong front legs to haul themselves up incredibly steep slopes and to brake when coming downhill. Their feet are specially adapted for clambering on loose rock and tiny ledges. A rim of hard, sharp hoof surrounds a flexible rubbery pad, which gives a good grip on even the most slippery rock or ice.

Mountain goats generally have a relaxed life, spending most of their time resting, dust-bathing, and feeding. They eat any plants that are available, browsing on trees and shrubs and nipping the tops off grasses and low herbs. During the summer they spend much of their time on high, rocky ledges, browsing on the small clumps of vegetation that manage to gain a foothold in the crevices or among loose rocks.

⊕ Mothers usually give birth on a high, narrow ledge, well out of reach of predators. Kids are able to walk soon after they are born and quickly learn to negotiate the craggy pathways.

Less often they descend to feed on lush alpine meadows. Females tend to have relatively stable home ranges, while males wander farther. As winter closes in and snow covers their feeding grounds, most mountain goats head for the lower slopes.

Predator Proof

Mountain goats seem unworried by predators. Most potential killers, such as coyotes and lynx, find it difficult to follow the goats up the high, rocky ledges. Small kids are most vulnerable, especially if they become trapped on the lower, less rugged slopes. They are also sometimes taken by golden eagles.

For most of the year males and females do not pay each other much attention. Animals usually feed alone or in small groups of a mother and her offspring. However, interactions can be aggressive when the animals fight for dominance or contest access to limited food supplies. The dominant animal varies, but generally nannies with kids rate highest and gain access to the best food.

Mountain goats do not fight head to head as sheep and true goats do. Instead, they stand side to side, each goat with its head toward the other's rear, tipping their heads to display the sharp-tipped horns. If posturing is not enough to settle disputes, pairs will spin around, trying to jab each other's rump and flanks with their spiky horns. Although the skin on these areas is thick, the short horns are formidable weapons, and animals are sometimes seriously or even fatally injured. Fighting is especially common in the rutting season (November to the end of December). Males scent-mark grass and tree branches by wiping them with oily secretions from glands at the base of their horns.

185

Pronghorn

Antilocapra americana

Pronghorns are the only survivors of a once successful North American antelope family. Their competitive lifestyle begins even before they are born.

PRONGHORNS ARE OFTEN REFERRED to as "antelope," since they are the only antelope species in the Americas. They are the fastest runners in North America, reaching speeds of over 55 miles per hour (88 km/h) in short sprints. When running fast, their long legs give them a huge stride of up to 20 feet (6 m). Their feet end in long, pointed, cloven hooves that are cushioned for running on rocks and hard ground.

Avoiding Predators

Speed is an adaptation to help pronghorns live relatively safely on open habitats, where there are few places to hide from predators. Being naturally curious, pronghorns will approach to inspect signs of movement in the distance, as do African antelope such as Thomson's gazelles. However, they are also wary creatures. If alarmed, they will run away, covering long distances before stopping to look back. They have excellent long-distance eyesight, their large protruding eyes giving them 360-degree vision. Long, black eyelashes act as sun visors.

Although similar in appearance and behavior to African antelope, pronghorns are only distantly related. Similarities are in fact due to both types of animals becoming adapted to similar habitats and lifestyles, a process known as convergent evolution. The pronghorn is the sole survivor of a once highly successful family of antelope species (the Antilocapridae) that roamed North America until the late Pleistocene epoch about 50,000 years ago. At that time they would have had to contend with large, fast predators, including North American lions, jaguars, and cheetahs. The necessity to escape speedily perhaps explains why pronghorns evolved to be such fast runners—a talent they have retained, although they need it less now.

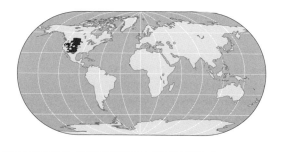

Common name
Pronghorn
(antelope)

Scientific name
*Antilocapra
americana*

Family Antilocapridae

Order Artiodactyla

Size Length head/body: 46–52 in (116–133 cm); tail length: 4–5.5 in (10–14 cm); height at shoulder: about 34 in (87 cm)

Weight Male 92.5–130 lb (42–59 kg); female 90–110 lb (41–50 kg)

Key features Long-legged antelope with stocky body; upperparts pale brown, white belly, flanks, throat, and rump; males have black face mask; single forward-pointing prong

Habits Active during day, with short feeding bouts at night; lives in single-sex herds for most of year; some populations migratory

Breeding Usually twins born after gestation period of 251 days. Weaned at 4–5 months; females sexually mature at 15–16 months, males at 2–3 years but breed later. May live 12 years in captivity, 9–10 in the wild

Voice Grunts and snorts; lambs bleat, males roar

Diet Forbs, shrubs, and grasses; often other plants such as cacti and crops

Habitat Rolling grassland and bush, especially dry sage brush country; open conifer forests

Distribution Western U.S., Canada, and some parts of northern Mexico

Status Population: over 1 million; IUCN various subspecies listed as Critically Endangered, Endangered, and Lower risk: conservation dependent; CITES I. Species as a whole no longer threatened

The horns of the pronghorn are unique. They have given rise to many arguments over the relationship between the pronghorn antelope and other horned ungulates. In the male the horns are large, about 13 to 16 inches (33 to 41 cm) long. They are backward-curving, and both have a single small, forward-pointing prong. Female horns are much smaller, often only 1.6 inches (4 cm) long, and do not have forward-pointing prongs. They may not be present at all. Each horn has a permanent, unbranched bony core, covered in a keratinous sheath that is shed every year after the rut. By renewing the sheath, the antelope can repair broken or frayed horn tips without discarding the whole thing (as deer do with their antlers).

During the rutting season females and males gather in huge herds. Females usually choose the male that has the territory richest in food. Although calves can walk within a few hours of birth, they do not have the stamina to run for long. To avoid predators such as coyotes, mothers hide their calves in long vegetation, only visiting them for a short period each day to groom and nurse them. At about three to six weeks the calves join a nursery herd with other mothers, calves, and yearlings. Males are sexually mature in their second year and leave their mother's group to join a bachelor herd. However, they rarely breed until they have a territory of their own. Females usually stay within their mother's herd.

Competitive Lifestyle

Life for pronghorns is highly competitive, with social status dictating intergroup and sexual relationships. Competition begins even before the calves are born. Four or more eggs may be fertilized at one time; and although all may implant in the uterus wall, usually only two survive. In the very early stages of their growth the fetuses at the top of the uterus produce long, hanging, tubular spikes that pierce and kill the embryos below. When calves join the nursery group, they jostle for top position, butting and chasing each other for the best feeding and resting sites. Larger calves—those born at the beginning of the season—usually win. The social rank they achieve as juveniles is often maintained for the rest of their lives.

⊖ *Male pronghorns use their horns in head-to-head fights, but only if their ritualized displays do not deter rivals.*

187

Common name Ord's kangaroo rat

Scientific name *Dipodomys ordii*

Family Heteromyidae

Order Rodentia

Size Length head/body: 4–4.5 in (10–11 cm); tail length: 5–6 in (13–15 cm)

Weight 1.3–1.9 oz (37–54 g)

Key features Small rat with large head, large black eyes, and rounded ears; hind legs and feet much longer than in front; very long tail with dark stripe along the top ending in tuft of long hairs; fur rich gold above, white below, with white band across tops of hind legs

Habits Nocturnal; lives alone in shallow burrow system

Breeding Up to 3 litters of 1–6 young born in spring or summer after gestation period of 29–30 days. Weaned at 3–4 weeks; sexually mature at 2 months. May live up to 10 years in captivity, usually many fewer in the wild

Voice Usually silent, sometimes uses foot drumming for communication

Diet Seeds and grains

Habitat Dry scrub, grassland, and desert

Distribution Southwestern Canada (Alberta and Saskatchewan) to northern-central Mexico

Status Population: abundant

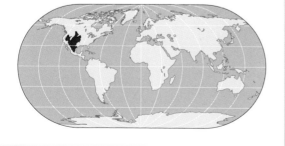

Ord's Kangaroo Rat

Dipodomys ordii

Like its close relatives the kangaroo mice and pocket mice, Ord's kangaroo rat has large cheek pouches for storing food. The feature is just one adaptation, demonstrated by this desert specialist to life in arid zones .

AS THEIR NAME SUGGESTS, KANGAROO rats have powerful back legs and move around by hopping. They can leap up to 6.5 feet (2 m), or well over 10 times their body length, in a single bound. The front legs are small by comparison, and the kangaroo rat uses them only for slow walking. Most of the time the front paws are used for sorting food and grooming.

Seed Stash

Ord's kangaroo rat is one of several species that survive in the deserts of western North America by feeding on the seeds that accumulate in the dry, sandy soils. Like pocket mice, kangaroo rats collect seeds in large, fur-lined cheek pouches and stash them away in underground burrows.

The burrows are shallow, presumably because the sandy soils are somewhat unstable. They contain nesting and food storage chambers, which are linked by interconnecting tunnels. A kangaroo rat's burrow is most definitely its castle, and the rather cute-looking animal will fight with surprising ferocity to defend its territory. Fights usually involve boxing with the front paws, kicking with the powerful hind legs, and scuffing sand into the opponent's face. Scent is known to play an important part in marking out a territory. A gland between the kangaroo rat's shoulders oozes an oily substance that the rat spreads liberally around its territory by wriggling in the sand. Kangaroo rats also use foot drumming as a means of alerting other individuals to their presence.

Survival Specialists

Kangaroo rats provide a textbook example of how to survive in arid conditions. Their success depends on a whole range of physiological and behavioral adaptations to help them keep cool and conserve water. They do not drink, and their food consists only of dry seeds and grains. In fact, their diet seems impossibly dry—scientists have estimated that the average kangaroo rat consumes only a tenth of the water it needs to survive. But the animal makes up the difference from water produced during the metabolism of its food. By creating and saving so-called "metabolic water," it compensates for the lack of moisture in its diet. It also makes use of dew on leaves at night.

Even so, making water takes energy, and so there is still precious little to waste. The kangaroo rat conserves what moisture it has by staying in its cool burrow during the day, therefore avoiding the need to sweat. Its ultraefficient kidneys produce urine that is four times more concentrated than a human's, and its droppings are almost completely dry. The kangaroo rat avoids losing water from its lungs by having a cool nose within which water vapor condenses before it can be breathed out.

Ord's kangaroo rat is one of the more widespread species of *Dipodomys*. Several of its close relatives are threatened by habitat destruction as people develop marginal land around deserts for agriculture or urban use.

⤴ *An Ord's kangaroo rat searches for seeds in the sand. Kangaroo rats are known to compete with smaller pocket mice for food where their territories overlap.*

Common name Northern pocket gopher
(western pocket gopher)

Scientific name *Thomomys talpoides*

Family   Geomyidae

Order   Rodentia

Size   Length head/body: 5–7 in (12–19 cm); tail
                 length: 1.5–3 in (4–8 cm)

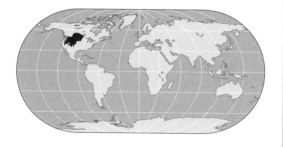

Weight 2–6 oz (57–170 g). Male up to
twice as heavy as female

Key features Robust body with short legs and large,
long-clawed feet; tail short with naked tip;
very thick neck and massive head, with small
eyes and ears; lips close behind prominent
incisor teeth; coat short and silky, any shade
of brown from near-black to creamy white

Habits   Solitary; aggressive; burrowing; mostly
                nocturnal

Breeding   Single litter of 1–10 (usually 3–5) young born
                in spring after gestation period of 18 days.
                Weaned at 40 days; sexually mature at 1 year.
                May live up to 4 years in the wild, probably
                similar in captivity, although not normally
                kept for very long

Voice   Generally silent

Diet   Roots, bulbs, tubers, stems, and leaves of
                various plants, including many crops

Habitat   Prairie grassland, forest, agricultural land, and
                anywhere else with soil suitable for burrowing

Distribution Southwestern Canada and western U.S.

Status   Population: abundant. Some subspecies may
                be threatened by habitat loss

Northern Pocket Gopher

Thomomys talpoides

*Pocket gophers are the most ancient living family of
burrowing rodents. They have spent over 36 million
years perfecting the art of excavation.*

THE NORTHERN POCKET GOPHER spends most of its
life in a private system of tunnels—some
shallow, some up to 10 feet (3 m) deep. The
deep burrows include sleeping and food storage
chambers, and the shallow ones are used for
foraging. Burrow entrances are usually sealed
with a plug of loose earth but are marked by a
distinctive fan-shaped heap of excavated soil. In
winter the tunnels extend out of the ground
and into the layer of overlying snow. The
gophers line their snow tunnels with soil, ridges
of which can be seen crisscrossing the
landscape after the spring thaw.

Most digging is done with the claws of the
front feet; but when the gopher comes across a
patch of compacted soil, it uses its front teeth

to gnaw its way through. The gopher's furry lips close behind the teeth so the animal can dig without getting dirt in its mouth. Even so, with such treatment and a diet of gritty roots, the gopher's incisor teeth are subjected to an immense amount of wear. Were it not for the fact that they grow continuously throughout the gopher's life, they would be eroded to nothing in the space of a few weeks.

Tails as Guides

Underground, pocket gophers can run forward and backward with almost equal speed. When going backward, the gopher uses its highly sensitive tail to feel its way along the tunnel. The tail tip is hairless and contains many nerves. The rest of the gopher's body is also sensitive to touch—the coat contains many special hairs that connect directly to nerves below the skin.

Male pocket gophers are considerably larger than females—a sure sign that size and strength are a factor in competing for mates. Gophers are aggressive animals and will fight over mates and territory. Most males carry serious scars from fighting, and the injuries inflicted are often fatal. Females will only permit males near their burrows in the mating season. During the rest of the year they will viciously repel intruders. The loose-fitting skin that allows gophers to turn easily in small spaces is equally useful in combat—it means that bites do not easily penetrate to the tissues beneath. Even when a gopher is held by the neck, it can still wriggle around and bite back.

Northern pocket gophers breed once a year. Maximum litter size is 10, but five is more normal. After two months the young gophers are fully weaned and independent, but face many dangers. Gophers dispersing from their mother's burrows are vulnerable to attack from carnivores, such as badgers and coyotes, and birds of prey. The northern pocket gopher is often persecuted because of its potential to do damage to crop plants. However, most control measures are only partially effective. Perhaps that is just as well because pocket gophers can also be highly beneficial, especially for grazing land. Their tunneling helps loosen and aerate the soil, improving drainage and encouraging the growth of herbaceous plants.

⊕ *Like their relatives the pocket mice, pocket gophers have large external cheek pouches in which they transport food. They do not appear to use the pockets for moving soil—that is shoved out of the tunnels using the feet, chin, and chest.*

Common name American beaver
(Canadian beaver)

Scientific name *Castor canadensis*

Family Castoridae

Order Rodentia

Size Length head/body: 31–47 in (80–120 cm); tail length: 10–20 in (25–50 cm)

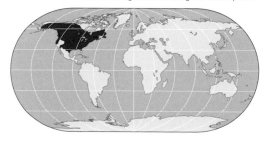

Weight 24–66 lb (11–30 kg)

Key features Robust body with short legs and large, webbed hind feet; tail scaly, flattened, and paddlelike; small eyes and ears; coat dense and waterproof, light to rich dark brown

Habits Lives in small territorial colonies of related animals; semiaquatic; fells small trees to build lodges and dams that are of great importance to wetland ecosystem; largely nocturnal

Breeding Single litter of 1–9 (usually 2–4) young born in spring after gestation period of 100–110 days. Weaned at 3 months; sexually mature at 18–24 months. May live over 24 years in captivity, up to 24 in the wild

Voice Hisses and grunts; also announces presence by slapping tail on water surface

Diet Aquatic plants such as water lilies and leaves; also bark, twigs, roots, and other woody tissues of waterside trees and shrubs

Habitat Lakes and streams among light woodland

Distribution Canada, Alaska, and much of contiguous U.S.; introduced to parts of Finland

Status Population: 6–12 million. Abundant—recovered well after serious decline due to excessive fur trapping in 18th and 19th centuries; regulated hunting still takes place

American Beaver

Castor canadensis

The industrious American beaver has helped shape the economic and ecological history of North America. Prodigious construction skills, a cooperative family life, and usefulness as a fur-bearing animal have given the rodent huge significance throughout its range.

AMERICAN BEAVERS ARE NATURALLY widespread, occurring from Alaska to northern Mexico. There are also thriving populations in Europe, especially in Finland, where the species was introduced following the decline of the native Eurasian beaver. Good beaver habitat consists of lightly wooded country dominated by species such as aspen, willow, and alder, all of which grow best in wet conditions. Water is vital to the beaver, which has a semiaquatic lifestyle.

Happy Families

Beaver colonies usually consist of one breeding pair along with their offspring of the past year or two. Beavers usually pair for life, and all members of the family help with chores such as maintaining the home (a lodge built of timber), watching over the babies, and gathering food. This kind of arrangement is close to the idealized human family, and it is one reason why beavers are regarded with affection by most people. Beaver family life is not without its trials, however, and young adults that have outstayed their welcome will eventually be driven out by their parents. Interactions with neighboring colonies are not necessarily amicable either: Members use scented heaps of dirt, twigs, and dung to mark out their territory and indulge in tail-slapping displays in the water to warn off intruders.

Where a beaver territory includes a suitable sheltered waterside bank, the family may set up home in a specially dug burrow. But if there is no natural site available, the resourceful beavers build one. First, they dam up a stream, creating

a large pool in which they can construct an island of timber and silt. Within the mound is a spacious living chamber, the entrance to which is underwater. These "lodges" are excellent places to bring up young beavers, since they are protected from most predators by the surrounding moat of cold water.

Construction Workers

Among the mammals the architectural and construction skills of beavers are second only to those of humans. Beaver dams and lodges are large structures, made mostly of timber—branches, logs, and sometimes whole trees up to 40 inches (100 cm) in diameter! That is unusual, however, and most trees felled are less than 10 inches (25 cm) thick. They are felled close to the water's edge by the beaver's huge gnawing teeth, then towed through the water and wedged firmly into place with dexterous front paws. Lodges, dams, and burrows are not the beaver's only feats of engineering. In order to transport enough timber for building, the beavers often have to excavate substantial canals through areas that are either too shallow or weed-choked to accommodate large floating branches without snagging.

Once built, beaver dams require constant maintenance and repair, especially during

← The North American beaver is perfectly adapted to its semiaquatic lifestyle. Its fur is warm and waterproof, and it can close its ears and nose while diving underwater.

Beaver Scent

The scents used by beavers to mark out their territory are produced in glands connected to the urinary tract. One of these scents is a substance unique to beavers and is known as castoreum. In the past castoreum was used to treat medical conditions including stomach cramps, ulcers, and various other aches and pains. The active ingredient that made these treatments effective is almost certainly salicin, a compound produced by willow trees. Salicin is the compound from which the common painkiller aspirin is derived. Beavers feeding on willow accumulate salicin in their body and use it in the production of scent. In recent times castoreum has become more commonly used as a base for perfumes.

spring, when many timbers are dislodged by fast-flowing melt water. A well-maintained dam will serve many generations of beavers, but they do not last indefinitely. Sooner or later the pool behind the dam becomes silted up, and the water will find an alternative route around the site, leaving the lodge high and dry. The resident beavers must then move on and begin again from scratch. The silted-up pools created in such a way eventually become willow thickets and lush meadows, replacing the woodland that once grew there. That offers opportunities for many species of insects, birds, and plants to live in an area where they would otherwise be absent. The beaver acts as a landscape architect, transforming the habitat. It is a good example of a keystone species on which many others depend for their survival.

Beavers do not hibernate, but in the north of their range they are seldom seen during winter. There is often a thick layer of ice and snow covering the pool, which effectively seals the beavers in. The silt in the thick lodge walls freezes solid, so that even if predators such as wolves and bears cross the ice, they are rarely able to break in. The beavers, however, can still come and go from the lodge by underwater entrances. During winter, when food is scarce, they can survive on plant material (mainly shoots and woody material) stored in special "caches" during the summer. The animals also have a reserve of fat stored in the tail, which helps them survive if spring comes late and food stocks run low.

During their winter confinement beavers live for several months without seeing true daylight. As result, their daily cycles of rest and activity are regulated by their own body clocks, rather than by the rising and setting of the sun. During these times they appear to

⊕ The beavers' home is a large pile of mud and branches sited on a riverbank or in the middle of a lake. It contains different rooms, with a living chamber above the water level and sometimes a dining area nearer the water.

switch from a regular 24-hour cycle to a longer one, something between 26 and 29 hours. Interestingly, the same happens with humans. People such as prisoners or experimental volunteers who live without natural light or artificial aids to telling the time develop a similar extended daily cycle and therefore often miscalculate the number of days they have been shut away.

Beaver Wars

The beaver's lustrous fur is soft, warm, and waterproof—qualities that are also greatly valued by humans. In the early colonial history of North America beaver trapping was so profitable that wars were fought over ownership of large areas of beaver habitat. Access to beaver skins was a major incentive to the exploration and opening up of the continent. Later, the fashion for felt top hats made from beaver fur encouraged still more trapping. Beavers were killed by the hundreds of thousands each year, and not surprisingly the population dwindled rapidly. Beavers are easy to

⊙ *Beavers only produce one small litter of kits each year. All members of the colony, which includes young of previous years, share tasks such as baby-sitting and providing the kits with solid food.*

find and are also easily trapped. By the early 20th century beavers had disappeared from much of their former range, and the species was in real danger of extinction. The loss of the beavers had enormous carry-over effects on entire wetland ecosystems. Without the beavers to build dams water drained rapidly from areas where it had formerly remained as pools. While excess water from heavy rainfall and spring thaws once spread gently over a wide area, in the absence of beavers it created raging torrents and flash floods.

Thankfully the danger was recognized in the nick of time, and legislation was put in place to preserve the remaining beaver stocks. Careful management has enabled numbers to recover, while allowing controlled trapping to continue. Beavers returned to much of their former range through recolonization, and other populations have been restocked artificially by bringing in beavers from elsewhere.

The reappearance of beavers is not always welcome, and there is an ongoing conflict between conservationists, trappers, and people who want to use the land for other purposes. Arable farmers claim that beaver activity harms their interests by flooding crop fields. Floods can also damage roads and other human infrastructure—one example of a situation in which commercial interests are at odds with ecological considerations. Establishing a compromise is one of the toughest challenges facing policymakers now and in the future.

⊙ *An American beaver building a dam. Although sometimes considered a nuisance by humans, dams in fact provide a natural filtration system that removes harmful impurities from the water. The large areas of wetlands that dams create also encourage greater biodiversity.*

Common name Gray squirrel
(American gray squirrel)

Scientific name *Sciurus carolinensis*

Family   Sciuridae

Order   Rodentia

Size   Length head/body: 9–11 in (24–29 cm); tail
length: 8–9 in (20–24 cm)

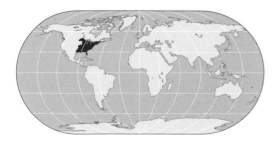

Weight 14–21 oz (400–600 g)

Key features Chunky squirrel with
variable silvery to dark-gray fur,
tinged with brown in summer; black
individuals or subgroups also occur; tail bushy
and fringed with white; ears rounded and
without hairy tips

Habits   Diurnal; tree-dwelling; also forages
extensively on the ground; solitary, but
nonterritorial; bold and inquisitive

Breeding   One or 2 litters of 1–7 young born
January–February and May–June after
gestation period of 44–45 days. Weaned at 8
weeks; females sexually mature at 8 months,
males at 10–11 months. May live up to 10
years in captivity, usually fewer in the wild

Voice   Churring and chattering sounds; screams in
distress

Diet   Seeds and nuts, especially acorns; also buds,
flowers, fruit, insects, and eggs

Habitat   Mixed deciduous woodland

Distribution Southeastern Canada and eastern U.S.;
introduced populations in Britain, Italy, and
South Africa

Status   Population: abundant

Gray Squirrel

Sciurus carolinensis

The gray squirrel is a native of North America, but thanks to widespread introductions it is now a familiar animal in parts of Europe too.

WITH ITS BRIGHT EYES, GREAT AGILITY, and inquisitive nature the gray squirrel is an appealing animal. It is also adaptable and can make the most of a wide variety of habitats. Its needs are simple: food and trees that are out of reach of predators in which to rear young. Its flexibility, along with a certain hardiness and tolerance of cold winter weather, has made the animal a big success in its native North America. Elsewhere, it thrives as a not altogether welcome alien.

Tree Houses

Gray squirrels will take advantage of holes in trees or hollow branches in which to build secure nests, where they shelter from the elements and raise their young. However, such ready-made nest sites are not very abundant, and in the absence of a tree hole most squirrels opt for self-built accommodation. Squirrels' nests (known as dreys) are highly distinctive. Slightly larger than a soccer ball, they are built of skillfully interwoven twigs nibbled from the same tree. Summer nests are built high in the tree tops, usually in a convenient fork and hidden from view by leafy branches. Winter dreys need to be a bit more robust in order to withstand buffeting winds and other winter weather. Dreys built in deciduous trees are inevitably rather exposed in winter, so the squirrels generally build close to the trunk where there is some shelter from the wind and rain and less movement in high winds.

Gray squirrels forage both in trees and on the ground. In natural habitats they feed mostly on nuts, pine cones, other seeds, and bark, but

⬆ *A female American gray snacks in Stanley Park, Vancouver. Squirrels are a common sight in parks and gardens across North America and Europe. Black phases like this one are often seen in the north of their range.*

The Bird-Feeder Challenge

For some people the gray squirrel's intelligence and its bold, inquisitive nature are part of its charm. For others the same characteristics result in a long-running battle of wits that more often than not the squirrels win. Garden bird-feeders provide an exceptionally good food source for any animal or bird that can reach them. Fat- and energy-rich foods such as nuts, seeds, and kitchen scraps are ample reward for a few hours spent solving the problem of access. The squirrels' ingenuity sometimes appears to know no bounds as time and again they figure out how to reach feeders hung in tricky places. Gray squirrels can climb bird tables and negotiate large overhangs with ease. They can run nimbly along wires and washing lines, haul up peanut feeders hung on string, bring down those suspended by wire, or split them open from the sides, spilling the nuts onto the ground. Some people gain a good deal of pleasure from devising new challenges for their garden squirrels, then sitting back and watching these smart rodents complete increasingly complicated obstacle courses for the reward of a handful of nuts.

they are happy to take food left out for birds and will regularly raid trashcans and unattended picnics. In parks and gardens they frequently become tame enough to take food from human hands. Like other squirrels, they sit up to feed and use their dexterous front paws to manipulate food. They lack a thumb, so they use two paws for grasping objects securely.

The reward makes all the effort worthwhile—a squirrel negotiates a garden bird feeder.

⊕ *A squirrel builds a nest in a hollow tree. Tree hollows are prime nesting sites for gray squirrels, providing shelter from the elements and a secure place to raise young.*

Most of the gray squirrel's natural food is seasonal, with late summer and fall being boom times. Squirrels take full advantage of the glut of nuts and acorns, and their appetite for mast is one of the main reasons for the massive overproduction of seeds by some trees. A single tree only has to produce one successful seed to reproduce itself, and yet the average oak produces thousands of acorns every year for centuries. In fact, millions of years of coexistence has resulted in mast trees and mast-eating animals (such as squirrels) adopting mutually beneficial strategies that help ensure their survival. A large tree produces huge crops of nuts, attracting squirrels to feed. But because the glut only lasts for a few weeks, the squirrels have learned to store excess nuts for the leaner times ahead. The gray squirrel buries nuts among the leaf litter and loose soil of the forest floor. It buries so many in a certain area that it does not need to remember the precise location of each nut. When the squirrel returns, all it has to do is dig and it will find food. Inevitably, some of the hidden nuts are never recovered, and they lie dormant in the soil, awaiting the chance to germinate.

Solitary Lives

Most female gray squirrels produce litters in spring and summer. The young squirrels spend their early lives in a secure nest, emerging at about six weeks old to begin exploring the treetops. Once they have left their mother's care, they are not especially sociable, although adults of both sexes will share nests in very cold weather. But for most of the year relationships are not quite so friendly. Males are especially pugnacious in the presence of an estrous female, and fights between rival suitors can result in bites, torn ears, and chunks of missing fur. Annoyance or aggression is signaled with growls and tail-wagging displays.

⊖ *A gray squirrel strips the bark from a sycamore tree in the hope of uncovering sappy tissue on which to feed. But its action can kill the tree.*

In their native North America gray and other tree squirrels are still common animals, although apparently less so than in the past. Their decline is due in part to man-made changes in their habitat and in particular to the

widespread loss of the American chestnut tree. In the 19th century squirrel populations could reach almost plague proportions during the summer and fall, when chestnuts were plentiful. If a hard winter followed, hordes of squirrels would perform mass migrations in search of more productive habitats. In the late 1960s millions of squirrels were killed every year, in part for their fur. The silvery coat of the gray squirrel is soft and warm, and was once fashionable. Squirrel hunting is now more closely regulated throughout most of the United States. Across the Atlantic in Britain the gray squirrel is doing so well that its total eradication, deemed by some to be desirable, seems virtually impossible.

Costly Imports

The gray squirrel's conquest of Britain began in the late 19th century, when the first animals were imported from the United States. The intention was to "improve" Britain's existing wildlife by adding new species to the fauna. This was in spite of the fact that Britain already had its own successful native squirrel—the highly arboreal red squirrel. However, gray squirrels are undeniably attractive animals with appealing behavioral characteristics and are easy to catch and transport.

The first batch of gray squirrels was released in England in 1876, and many more followed over the next 50 years. The species did exceptionally well; and by 1930, when the less desirable consequences of their spread were realized, it was too late to reverse the trend. As the gray population boomed, the red squirrel entered a dramatic decline. Despite vigorous attempts to control their number, gray squirrels thrived, and by 1960 they had colonized all but the far north of England and Scotland. As the gray squirrel spread, the native species was displaced and steadily died out over large areas. Gray squirrels have caused similar problems in Ireland and have now been released in northern Italy. They have also been taken to South Africa, where they are responsible for significant damage to native and plantation trees.

Common name Eastern chipmunk

Scientific name *Tamias striatus*

Family Sciuridae

Order Rodentia

Size Length head/body: 5.5–7 in (14–17 cm); tail length: 3–5 in (8–12 cm)

Weight 3–5 oz (85–140 g)

Key features Large chipmunk with reddish-brown coat, fading to cream on belly, with 5 black stripes along back separated alternately by brown-and-white fur; large internal cheek pouches; bottlebrush tail covered in short hair; ears rounded, eyes large and bright; face also has striped markings

Habits Diurnal; solitary; territorial; digs extensive burrows—may stay there in winter, but does not hibernate

Breeding One or 2 litters of 1–9 (usually 3–5) young born in spring and summer after gestation period of 31 days. Weaned at 6 weeks; sexually mature at 10–12 months. May live up to 8 years in captivity, 3 in the wild

Voice High-pitched chirping calls

Diet Nuts, seeds, acorns, fungi, fruit, and crop plants; occasionally insects, birds' eggs, and baby mice

Habitat Lightly wooded land with warm, dry soils and rocky crevices in which to hide

Distribution Eastern U.S. and southeastern Canada

Status Population: abundant. Threatened by persecution and habitat loss in some agricultural areas

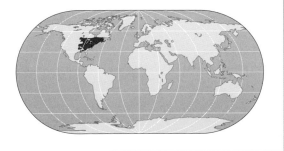

Eastern Chipmunk

Tamias striatus

The eastern chipmunk is one of the most widespread and best known members of the genus Tamias. *Its name means "treasurer"—a reference to the chipmunk's compulsive hoarding of nuts and seeds to help see it through the winter months.*

EASTERN CHIPMUNKS ARE LARGER than the other 24 species of *Tamias* and are generally found in areas of relatively light deciduous woodland. They are especially fond of places with broken rocky ground or old stone walls, since they provide plenty of secure nooks and crannies in which to hide from predators.

Excavation Skills

Chipmunks of this species are accomplished diggers, and their burrows can extend up to 30 feet (9 m) underground. The entrance to the burrow is usually well hidden at the base of a rock or fallen log and often among a layer of loose leaves. Unlike some other ground-dwelling squirrels, the eastern chipmunk carries away soil excavated from its burrow for disposal so there is no telltale mound of spoil to give away its whereabouts. The soil is carried in two capacious cheek pouches, which open inside the chipmunk's mouth. The same pouches are also used for transporting food such as acorns and nuts. When full, the pouches are almost as large as the chipmunk's skull.

Stored food is a vital resource that enables chipmunks to survive the winter. Unlike their Siberian cousin *Tamias sibiricus*, American chipmunks do not enter full winter hibernation, and they need to continue feeding even when their homes are buried deep in snow. They therefore spend the fall gathering enough food

⊕ The eastern chipmunk has large cheek pouches in which to transport food items, as well as soil excavated from its burrow.

to see them through several months of enforced inactivity. They spend much of the winter asleep in their burrows, but wake regularly to feed and will emerge into the open during short periods of warmer weather. When spring finally arrives, they launch almost immediately into breeding activity and in good years manage to fit in two breeding seasons before the end of summer.

Belligerent Behavior

Adult eastern chipmunks are territorial and aggressive. The high-pitched "chip, chip, chip" call for which they are named is used to advertise boundaries and warn off intruders. Confrontations are common because, while core territories are small—usually no more than a 50-foot (15-m) radius around the burrow entrance—each chipmunk will regularly venture up to 150 feet (45 m) from its burrow to forage. While out and about it often meets with its neighbors.

Aggression reaches a peak during the mating season, when males gather in the territory of an estrous female to compete for her attention. However, the female still has the final word about which male fathers her offspring, and unworthy suitors are unceremoniously rejected. The young (usually numbering three to five) are born in the female's burrow and emerge into the open at about six weeks of age. At eight weeks they can fend for themselves and will disperse into the local area, usually settling nearby.

Many people enjoy watching chipmunks in parks and gardens throughout the animals' range. The chipmunks are bold, inquisitive, and charismatic and can be tamed fairly easily. However, despite their obvious appeal, they are also capable of causing serious damage to crops, garden plants, and other property. In some cases their burrows can destabilize fence posts and even undermine the foundations of buildings.

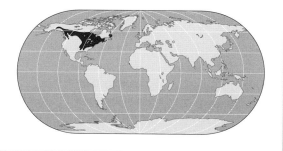

Woodchuck

Marmota monax

Also known as the groundhog or whistlepig owing to its high-pitched alarm call, the woodchuck is one of North America's best-known rodents.

Common name Woodchuck (groundhog, whistlepig)

Scientific name *Marmota monax*

Family Sciuridae

Order Rodentia

Size Length head/body: 16–26 in (41–66 cm); tail length: 4–10 in (10–25 cm)

Weight 4.5–13 lb (2–6 kg)

Key features Chunky, short-legged, short-tailed rodent; dense, woolly, reddish-brown fur with white-tipped guard hairs; head darker, with small eyes and small, rounded ears; short, hairy tail

Habits Diurnal; terrestrial; burrowing; hibernates in winter

Breeding Single litter of 1–9 (usually 4 or 5) born in spring after gestation period of 30–32 days. Weaned at 6 weeks; sexually mature at 2 years. May live up to 15 years in captivity, 6 in the wild

Voice Loud whistling alarm call; also makes chattering, squealing, and barking noises; grinds teeth when agitated

Diet Fresh grasses and leaves of other plants; also seeds and fruit

Habitat Prairie, grasslands, and road edges

Distribution North America: Alaska to Idaho in the west, Newfoundland to Alabama and Arkansas in the east

Status Population: abundant. Has expanded range and population in some areas developed for agriculture, but hunting for fur and persecution as a pest have led to significant declines in others

THE AMERICAN WOODCHUCK is a species of marmot—a kind of large, ground-dwelling, burrowing squirrel. It is a highly proficient digger: Its legs are short, powerful, and bear strong claws to help loosen packed soil, which is then shoved aside by the soles of the woodchuck's flat feet.

Woodchuck burrows are extensive: Some extend horizontally as much as 50 feet (15 m) and have four or five entrances. The burrows are a nuisance to farmers because they cause damage to plowing machinery, undermine the foundations of buildings, and pose a threat to livestock, which can stumble into the holes. The woodchuck tends to use different dens in different seasons. In summer it lives and breeds in burrows dug in open ground and marked by a heap of fresh dirt at the entrance. In winter it retreats to a less conspicuous burrow, usually dug at the base of a tree. The roots may help support the tunnel during wet weather.

Homeless Youngsters

In prime habitat woodchuck population densities can exceed 37 per square mile (15 per sq. km). Individual home ranges vary with habitat and season. They can be as small as half an acre (0.2 ha) for breeding females or as large as 7.5 acres (3 ha) for males, whose ranges usually overlap those of several females. The woodchuck is the least social of the marmots, and the ranges of males and females tend not to overlap with those of other animals of the same sex. Territories are advertised using scent from three large glands under the tail, and intruders are aggressively driven away. Females can also be ruthless in expelling their own offspring from their territory as soon as they are able to fend for themselves—at about six weeks old. Daughters are sometimes permitted to stay over the winter, but only if there is a good supply of food.

Unlike most other squirrels, woodchucks favor green leaves and shoots over energy-rich seed heads and nuts. Their relatively low fat and calorie content mean that woodchucks must eat a lot to sustain themselves, sometimes over 1 pound (0.4 kg) of vegetation every day. Their large appetites make woodchucks very unpopular with farmers—the hungry rodents can ruin crops such as alfalfa, corn, and oats.

Woodchucks must feed well in spring and summer because in winter they enter deep hibernation. The duration varies, but the animal usually emerges sometime in March or April. American folklore tells that if the woodchuck emerges on February 2 and sees its own shadow (i.e., if the sun is shining), there are six more weeks of wintry weather to come; if the day is cloudy, then spring is not far away. The precise origins of the story are obscure, but thanks to the 1993 movie of the same name millions of people have heard of "Groundhog Day" even if they do not know what it means!

⊕ Two woodchucks in Minnesota perch on a plank of wood. The rodents are also known as groundhogs, and February 2 is "Groundhog Day."

203

Common name Thirteen-lined ground squirrel (thirteen-lined suslik)

Scientific name *Spermophilus tridecemlineatus* (*Citellus tridecemlineatus*)

Family Sciuridae

Order Rodentia

Size Length head/body: 4–7 in (11–18 cm); tail length: 2.5–5 in (6–13 cm)

Weight 4–5 oz (110–140 g). Weight can almost double prior to hibernation

Key features Compact body with short legs and furry tail up to about half length of body; coat strikingly marked with 13 alternating dark and pale stripes; the dark stripes are patterned with white spots

Habits Diurnal; burrowing; hibernates for up to 8 months a year

Breeding One (occasionally 2) litters of 2–13 young born in summer after gestation period of 28 days. Weaned at 6 weeks; sexually mature at 9–10 months. Females may live up to 11 years in captivity, similar in the wild; males about half as long

Voice Twittering chirps and chattering sounds; growls and piercing whistle used as alarm call

Diet Grasses and other herbs; seeds, insects (especially beetles, grasshoppers, and caterpillars); also eggs, baby mice, and birds

Habitat Grass prairie and farmland

Distribution Southern Canada (Alberta) south to Ohio and southern Texas

Status Population: abundant

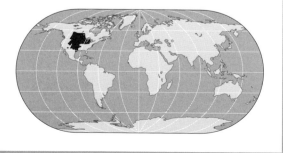

Thirteen-Lined Ground Squirrel

Spermophilus tridecemlineatus

The boldly patterned thirteen-lined ground squirrel is one of about three dozen American and Eurasian species belonging to the genus Spermophilus, *which literally translated means "seed-lover."*

THE GROUND SQUIRRELS' FONDNESS for seeds and grains makes many species unpopular with farmers. The thirteen-lined ground squirrel redeems itself somewhat by also eating large quantities of insects such as beetles and grasshoppers, which can themselves be serious agricultural pests. Unlike some species of *Spermophilus*, which have declined as their habitat has been taken over by humans, the thirteen-lined species thrives in areas where its natural prairie habitat is modified to grow crops. It also favors areas where forests are felled to make way for agriculture.

Strikingly Marked

The classification of the ground squirrels is tricky because many species appear similar. However, there is no mistaking the thirteen-lined ground squirrel, which is much more strikingly marked than its close relatives. From the age of about three weeks individual squirrels are marked with 13 bold black and buffy-white stripes down the back. The black stripes are further adorned with white spots, making the thirteen-lined ground squirrel one of the world's most decorative and attractive rodent species.

Thirteen-lined ground squirrels are active by day, but only during the summer months. In the fall, when food becomes scarce, they enter a long hibernation of up to 240 days. They sleep in grass-lined chambers excavated at the end of tunnels 16 to 20 feet (5 to 6 m) in length. The same tunnels are used for shelter during the

the four or five
months it is active
as fat is laid down for
winter. Individuals also
collect seeds in their
cheek pouches and cache
them away underground so that
they are guaranteed a good
breakfast when they wake up, even if
spring is late coming and food still scarce.

Race Against Time

With such short periods of activity each year it
is vital that breeding efforts start as soon as
possible after the squirrels emerge from their
winter sleep. Males begin seeking mates about
two weeks after waking, and the first litters are
born four weeks later. Experienced females tend
to have large families of up to 12 young, while
first-time mothers bear smaller litters of three or
four. There is hardly time for the offspring to
wean and fatten up before they need to enter
hibernation. Late summer sees a race against
time for youngsters trying to build up reserves
to survive the winter. When to stop feeding and
enter hibernation is a difficult judgment. Every
day that the squirrel is active costs it more in
terms of energy. Yet if it is underweight, it must
continue foraging for foods that get harder to
find as winter approaches.

active season and
for rearing young. The
entrances are usually
concealed at the base of rocks or
fence posts, and ground squirrel
communities are often strung out along the
fence lines that cross open farmland. Unlike
many other species of *Spermophilus*, they are
neither particularly social nor territorial,
although each adult likes to keep a burrow for
its own personal use.

Hibernation removes the need to eat
during winter, since the sleeping animals live off
reserves of fat accumulated during the summer:
A healthy adult doubles its body weight during

⊕ *The thirteen-lined
ground squirrel is small
and easily hidden among
grass, where its
patterned coat breaks
up its outline.*

Common name Black-tailed
prairie dog (plains prairie dog)

Scientific name *Cynomys ludovicianus*

Family Sciuridae

Order Rodentia

Size Length head/body: 10–12 in (26–31 cm); tail
length: 3–4 in (7–9.5 cm). Female about 10%
smaller than male

Weight 20–53 oz (575–1,500 g)

Key features Sturdily built squirrel with short legs and
short tail with black tip; coat buffy gray

Habits Diurnal and fossorial; highly social but also
territorial; does not hibernate

Breeding Single litter of 1–8 (usually 3–5) young born
in spring after gestation period of 34–37
days. Weaned at 5–7 weeks; sexually mature
at 2 years. May live up to 8 years in captivity,
5 in the wild

Voice Various barks, squeaks, and soft churring
sounds

Diet Grasses and herbs

Habitat Open short-grass plains and prairies; also
pastureland

Distribution Great Plains of North America from
southern Canada (Saskatchewan) to northern
Mexico

Status Population: more than 1 million; IUCN Lower
Risk: near threatened. Has declined due to
habitat modification and persecution

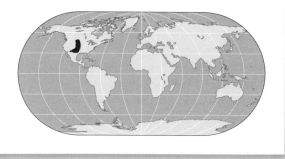

Black-Tailed Prairie Dog

Cynomys ludovicianus

*Seemingly empty expanses of land can
be home to hundreds of thousands
of prairie dogs living in huge, highly
organized rodent cities under
the ground.*

THERE IS NOTHING PARTICULARLY doglike about the
appearance of the black-tailed prairie dog and
its close relatives: They are in fact members of
the squirrel family. However, they are highly
social animals, and some of their vocalizations
sound a bit like those of small dogs. The most
distinctive call is known as the "jump-yip"—a
sharp double bark usually given as a territorial
warning and accompanied by a sudden half
jump in which the animal rears up on its hind
legs and flings its "arms" in the air as though
trying to shoo the intruder away.

Subterranean Towns

Black-tailed prairie dogs live in huge, highly
organized colonies known as "towns." Like
human settlements, prairie dog towns vary in
size and number of inhabitants. Most cover
about 250 acres (100 ha), but they can be
much larger. The largest town ever recorded
was reputed to cover almost 25,000 square
miles (64,750 sq. km) and was home to about
400 million black-tailed prairie dogs—the same
as the human populations of the United States
and Japan combined!

Within a town there are distinct
neighborhoods known as "wards." They can be
any size and are usually defined by obvious
natural features like banks or clumps of
vegetation. Within each ward there are a
number of highly territorial family groups called
"coteries." They contain eight or nine animals
on average, but sometimes more than 20. The
coterie is the basic social unit of the prairie dog,
and the animals within such a group know each

other well and share close blood ties. Social bonds between the members of a coterie are constantly reinforced by kissing, nuzzling, and mutual grooming and by a variety of vocalizations. The burrows of animals in the same coterie may be interconnected, but tunnels do not cross coterie borders.

Within a coterie there is usually one breeding male and three or four breeding females. The other members of the group are juveniles from the last two breeding years. Young male prairie dogs disperse from the group before they reach maturity at two years of age, while females remain to breed. However, they do not mate with their own fathers because adult males usually move on before their daughters are able to breed. In large coteries dominance and breeding rights may be shared by two males, often brothers.

Homicidal Females

Female prairie dogs can only breed once a year. Breeding is synchronized so all young are born at more or less the same time. During the breeding season (spring) females are far less sociable than usual, and they begin to exclude other members of the coterie from their breeding burrow. They are known to attack and kill the young of other females. Such brutality among closely related mammals is very unusual and has not yet been fully explained. Perhaps the females are attempting to reduce overcrowding and pressure on resources, giving their own young a better chance. Or it may simply be that they need to boost their own strength by eating meat—something they do not normally do. In contrast to the breeding females, the adult males in the coterie treat all young with equal benevolence.

Unlike most other ground squirrels, prairie dogs do not store food for winter. Therefore, they have two choices when it comes to surviving the harsh weather: Either

⊖ *A black-tailed prairie dog feeding on plant material. Unlike many species of ground squirrel, the black-tailed prairie dog does not hibernate and continues to forage for food throughout the winter.*

A Home on the Prairie

There is far more to a prairie dog burrow than a simple tunnel in the ground. The burrows are built to precise specifications. Not only are they the perfect size and shape for their owners, but they are also designed with emergency entrances, flood defenses, and built-in air conditioning. Tunnels leading to sleeping chambers are about 4 to 6 inches (10 to 15 cm) in diameter and may extend as far as 100 feet (30 m) underground. Chambers are 12 to 18 inches (30 to 45 cm) across. There is normally more than one entrance, sometimes as many as five or six, and they drop vertically down into the tunnels for easy access. There is usually a shelf or chamber just inside the entrance to make climbing out easier. The soil that is excavated from the burrow is heaped around the entrance, where the prairie dog carefully compacts it and creates a conical mound, like a volcano, with the burrow entrance in the central crater. Having a raised entrance prevents rainwater from running into the hole, and the burrows stay remarkably clean and dry all year round. Because the burrows can be anything from 3 to 16 feet (1 to 5 m) deep, there is a danger of air trapped inside becoming very stale. But having different-sized mounds at each entrance means that when a breeze blows over the ground, air moves faster over some burrow entrances than others. That creates a gentle flow of air in the burrow system and keeps it well ventilated. The burrow is a complicated piece of engineering to understand, and yet the black-tailed prairie dog instinctively knows how to build it.

they remain active and continue to forage for food, or they must hibernate in order to save energy until spring. While the Utah and white-tailed prairie dogs do hibernate, the black-tailed species remains more or less active through the winter. It may remain dormant in its burrow during very cold weather, but rouses regularly to feed and never enters the deep, torpid slumber of true hibernation.

Prairie dogs eat mostly grass and herbaceous vegetation. They forage in a systematic fashion, nibbling the turf in one part of their range very short before moving on and leaving the close-cropped turf for a while to regrow. Regular clipping quickly eliminates tall vegetation and encourages the growth of low, vigorous plants. Prairie dogs never let the grass get long enough to provide cover for larger animals, and so predators such as coyotes or badgers find it impossible to approach unnoticed. Group security is further improved by the members of a coterie taking turns standing guard, often using one of the large burrow entrance mounds as a lookout post. However, there is little doubt that their greatest threat comes from humans.

Rise and Fall

Early on, the spread of European settlers across the prairies gave a significant boost to the prairie dog population. Pasture created for livestock turned out to be ideal prairie dog habitat and the ranchers obligingly killed or drove away many natural predators. By 1900 there were an estimated 5 billion black-tailed prairie dogs in the United States. However, the species soon became regarded as a menace. Not only do prairie dogs eat a lot of grass, their burrows destabilize the ground, creating a hazard to livestock and a nuisance to farm machinery. In the early part of the 20th century prairie dogs were poisoned by the millions and their towns were plowed up and destroyed. The population today is less than 0.1 percent of what it was 100 years ago, and the majority of surviving colonies live in protected areas such as national parks. Some of the species' close relatives are even more scarce.

⊕ Two prairie dogs greet each other by pressing nose to nose. Prairie dogs are a social species whose bonds are reinforced by interactions such as nuzzling, "kissing," and grooming.

⊖ A black-tailed prairie dog family in Colorado. The family, or "coterie," is the basic social unit of prairie dogs, members of which know each other extremely well.

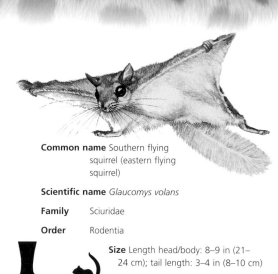

Common name Southern flying squirrel (eastern flying squirrel)

Scientific name *Glaucomys volans*

Family Sciuridae

Order Rodentia

Size Length head/body: 8–9 in (21–24 cm); tail length: 3–4 in (8–10 cm)
Weight 1.7–4 oz (50–120 g)

Key features Small silvery-gray squirrel with bushy but flattened tail and furry gliding membrane stretching from wrists to ankles; head large with big ears and huge black eyes

Habits Nocturnal, social, and gregarious; arboreal; hoards food—does not hibernate; "flies" by gliding on flaps of skin stretched between front and back limbs

Breeding One or 2 litters of 1–6 (usually 2–4) young born in spring and summer after gestation period of 40 days. Weaned at 8–9 weeks; sexually mature at 9 months. May live up to 14 years in captivity, considerably fewer in the wild

Voice Chirping and squeaking calls, many of which are too high-pitched for humans to hear

Diet Nuts, acorns, bark, fungi, fruit, and lichen; occasionally eats insects and meat

Habitat Woodland

Distribution Southeastern Canada, eastern U.S., and Central America south to Honduras

Status Population: abundant. Has declined in some areas due to deforestation, but generally secure; may be responsible for the decline of the related but much rarer northern flying squirrel in some areas

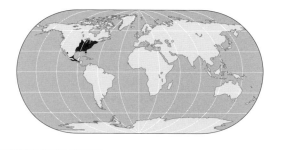

Southern Flying Squirrel
Glaucomys volans

The southern flying squirrel is one of about 30 species of tree-dwelling rodents that have solved the problem of traveling from tree to tree by taking to the air.

AT FIRST GLANCE BABY FLYING squirrels look similar to most other infant rodents—tiny, pink, hairless, and with their eyes and ears sealed over with skin. Their legs are rather long, but closer examination reveals something far more unusual: The babies appear to be wearing skins that are many sizes too big for them. A large fold of skin forms a double membrane that links the wrist to the ankle along each flank. As the baby grows, the extra skin thickens and becomes furred along with the rest of the body. At six or seven weeks of age the young squirrels first emerge from their nest in a tree hole, nest box, or attic. At about eight weeks the purpose of the extra skin suddenly becomes clear: The squirrels can be seen taking to the air like miniature stunt parachutists.

Skillful Pilots

A flying squirrel's "flights" are in fact carefully controlled glides. Before launching itself with a powerful thrust of its back legs, the squirrel examines its distant landing site from several angles, tilting its head this way and that to help gauge distance and trajectory. Once airborne, it opens its gliding membrane by spreading its legs wide. This greatly slows its descent and allows it to travel up to 3 feet (1 m) horizontally for every foot of vertical descent. So from a 60-foot (18-m) perch a squirrel can "fly" up to 150 feet (50 m) to a point well above ground level in another tree. The flattened tail acts as a rudder; and by changing the shape of its parachute, the squirrel can exert fine control over its glide, even swerving in midair. On landing, the squirrel instantly scurries around the tree trunk in case its glide was noticed by the sharp eyes of an owl.

The southern flying squirrel is the smaller of two closely related species living in North America. However, it is more common than its relative the northern flying squirrel (*Glaucomys sabrinus*) and, despite its diminutive size, appears to be dominant in areas where the distribution of the two species overlaps.

Winter Provisions

Most of the year southern flying squirrels are sociable. They actively seek out each other's company—especially in cold winters, when up to 50 squirrels may be found huddled together in a secure nest. The animals do not hibernate, but they will remain inactive during bad weather. Flying squirrels feed on nuts and acorns, which they gather in large quantities during the fall. They hide the food in nest holes and other nooks and crannies to provide a winter larder.

Male flying squirrels share home ranges all year round. However, during the breeding season females establish a personal territory, which males visit to compete for the right to mate. Females are capable of rearing two litters a year, but that only happens if spring comes relatively early and is more common in southern parts of the species' range.

⊖ *A southern flying squirrel prepares to land on the trunk of a tree. At the last moment it swings its back legs forward so that all four feet land together.*

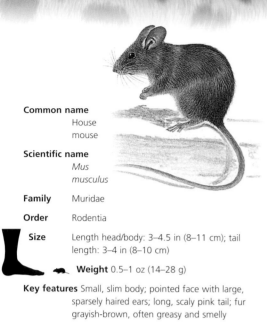

Common name	House mouse
Scientific name	*Mus musculus*
Family	Muridae
Order	Rodentia
Size	Length head/body: 3–4.5 in (8–11 cm); tail length: 3–4 in (8–10 cm)
Weight	0.5–1 oz (14–28 g)
Key features	Small, slim body; pointed face with large, sparsely haired ears; long, scaly pink tail; fur grayish-brown, often greasy and smelly
Habits	Generally nocturnal; often aggressive; excellent climber, also swims well; lives wild and in association with people
Breeding	Up to 14 litters of 4–10 young born at any time of year after gestation period of 19–21 days (more if female is suckling previous litter). Weaned at 3 weeks; sexually mature at 6 weeks. May live up to 6 years in captivity, 2 in the wild
Voice	Squeaks
Diet	Omnivorous: almost anything of plant or animal origin, including leather, wax, cloth, soap, and paper; also chews man-made materials such as plastics and synthetic fabrics
Habitat	Farms, food supplies, fields, and houses
Distribution	Almost worldwide
Status	Population: billions. Less common than previously in many developed countries due to intensive pest control and mouse-proof buildings

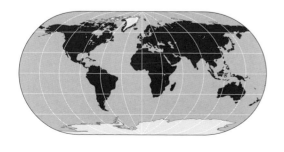

House Mouse

Mus musculus

The house mouse is a familiar animal all over the world. Ecologically it is a "Jack-of-all-trades," and its generalist lifestyle has allowed it to make a living anywhere that people can.

TRADITIONAL STORIES AND CARTOON animations would have us believe that the sight of a house mouse is enough to send housewives (and their husbands, too) shrieking onto the nearest stool. Despite their reputation, house mice are not really that exciting to look at. In fact, they are drab little animals, with coarse, grayish-brown fur, which is almost always slightly greasy, not sleek and dry like that of a wood mouse. The house mouse smells musty and rather unpleasant, and it moves like greased lightning, but it is not entirely without charm. It has bright black eyes, large charismatic ears, and an inquisitive, twitchy nose. All in all, it is difficult to see why anyone should be frightened of one.

Association with Humans

The house mouse first appeared on the steppes of Central Asia, but is now one of the world's most widespread mammals. It lives wild in deserts, swamps, hedgerows, and forests, from the sweltering tropics to windswept islands. However, the house mouse's big break in terms of conquering the world was undoubtedly its association with humans. House mice can live anywhere that people can, from cities to Antarctic research stations, coal mines to mountain huts. They can be found in subways and public buildings, and are even known to take up residency in the finest hotels.

House mice do best in traditionally managed farmland, where large quantities of crops or animal feed are stored in barns. In such a place mouse populations can increase spectacularly. When the owner of one such farm in Australia began a poisoning campaign to control the mice, he awoke next morning to find almost 30,000 corpses on his doorstep.

⬆ *A house mouse feeding on blackberries growing in a hedgerow. The diet of house mice is exceptionally varied and can include such items as wax and soap.*

Mouse Proof

For most people living in towns and cities in the developed world, the scourge of the house mouse is no longer really a reality. Modern building technology and food storage facilities mean that most homes are pretty much mouse proof, and mouse problems are easily dealt with by poisoning. It is known that in some places mice have become resistant to certain poisons that are used to control them, but there are more being developed all the time. As a result, mouse infestations are now little more than a nuisance in many parts of the world. Mice have a way of getting into all kinds of unexpected places. They wreak havoc by chewing fabrics, paperwork, and even electrical cables.

Singing and Dancing Mice

House mice have been specially bred to make pets and laboratory animals. There are many different strains, some of which are distinctive enough to have been given names. "Shaker" and "waltzing" mice actually have something wrong with the part of their nervous system that controls balance, meaning they cannot walk properly. Shaker mice weave and wobble as they walk, while waltzers seem to run in constant dizzying circles, going nowhere fast. Under normal circumstances such defects would be catastrophic to individuals, but sometimes life is so easy for house mice that afflicted animals are still able to get by and reproduce.

So-called "singing mice" habitually make twittering sounds that can be heard from some distance away. Drawing attention to themselves would be disastrous for most small animals—and no doubt any wild singing mice would meet a sticky end if overheard by a cat or weasel. However, in the relative safety of human habitations the mice can often breed fast enough that some survive to pass on their noisy genes. Singing mice have also been specially bred as novelty pets.

House Mouse Behavior

The social behavior of house mice varies depending on their circumstances. At high population densities they do not always defend territories, but often when living alongside humans they become distinctly territorial. Male mice are more aggressive than females and will fight viciously in order to establish dominance. A male shares his territory with several females, with whom he mates. The females have a dominance hierarchy of their own, but it is more peacefully maintained. Both females and young mice help the dominant male repel intruders from the family territory.

House mice mark their territory with smelly droppings and urine. Sometimes the feces accumulate as little stinking pyramids where generations of mice have deposited their messages in the same place. The copious droppings and urine contaminate food, hay, and bedding materials. The mice also gnaw packages, wood, wires, and even lead shot, greatly adding to the economic losses they cause by simply eating stored food.

Apart from its extraordinary adaptability and ability to feed on almost anything, a large

⊕ Two-week-old mice huddle together in a nest. There are usually four to 10 young per litter and up to 14 litters in a year.

part of the house mouse's success is due to its reproductive strategy. Female house mice are able to breed when they are just six weeks old. From then on they can produce increasingly large litters roughly once a month for about a year. The offspring mature quickly, but tend not to breed themselves until they have dispersed far enough away from their mother. If they do not move on, they remain under the influence of special chemicals (pheromones) that inhibit ovulation. Inbreeding can lead to birth defects and weakened immune systems, but the pheromones produced by the mother prevent young females from breeding with their own father or other closely related males. It is the combination of rapid breeding and dispersal that enables the house mouse to exploit new opportunities and spread so quickly.

Once established in a breeding range, house mice can be highly sedentary. Studies of mice in grain silos have shown that many individuals never wander more than a couple of yards from their nests. Everything they need is easily found in a small area, and they do not even need to drink where there is ample water in their food. Away from human habitation, house mice are a little more venturesome, traveling several hundred yards in a night and sometimes dispersing more than a mile.

All house mice are incredibly agile. They can climb branches, ropes,

walls, and cables, and they are able to squeeze through minute gaps and under doors. They can also run fast, about 8 miles an hour (13 km/h), faster than the average human jogger several hundred times their size.

Local Varieties

Although they are descended from a relatively recent shared ancestor, house mice in different parts of the world show distinct regional variations. Many local varieties are now classified as subspecies of *Mus musculus.*

Some of the differences between the subspecies can be used to trace the routes by which the house mouse has conquered the world. The closest thing to the ancestor of all house mice is thought to be a subspecies called *Mus musculus wagneri*, which originated in the steppe county between the Black and Caspian Seas. Interestingly, the area also contains evidence of the earliest arable farming: Its people started cultivating

cereal crops several thousand years ago. No doubt the mice took advantage of stored grain, thus forming a relationship with humans that has continued to this day. Mice were unwittingly transported with grain being traded east and west, and by 4,000 years ago they had spread to Europe and North Africa. By 1200 B.C. there were house mice in Britain, and by medieval times there were two distinct subspecies in northern and southern Europe.

The southern subspecies (*Mus musculus brevirostris*) made it to South and Central America and California with the Spanish and Portuguese conquistadors. At the same time, the northern variety (*Mus musculus musculus*) was colonizing North America along with the Pilgrim Fathers. Nowadays a large proportion of western Europe has been taken over by another subspecies (*Mus musculus domesticus*), which is characterized by a longer tail.

⬇ *A house mouse feeds on grain stored in a sack. Wherever cereal crops are cultivated, house mice also thrive. The species' relationship with humans can be traced back to the earliest arable farming, several thousand years ago.*

Common name Brown rat (common rat, Norway rat)

Scientific name *Rattus norvegicus*

Family Muridae

Order Rodentia

Size Length head/body: 9–11 in (22–29 cm); tail length: 7–9 in (17–23 cm)

Weight 9–28 oz (255–790 g)

Key features Typical rat with short legs, longish fingers and toes, and pointed face; ears pink and prominent; scaly tail noticeably shorter than head and body; fur dull grayish-brown, fading to white or pale gray on belly

Habits Generally nocturnal; social; cautious at first but can become bold; climbs and swims well

Breeding Up to 12 litters of 1–22 (usually 8 or 9) young born at any time of year (but mostly in spring and summer) after gestation period of 21–26 days. Weaned at 3 weeks; sexually mature at 2–3 months. May live up to 6 years in captivity, 3 in the wild

Voice Loud squeaks when frightened or angry

Diet Anything edible, including fruit, grain, meat, eggs, wax, and soap; will catch and kill other small animals

Habitat Almost anywhere food can be found

Distribution Worldwide in association with humans; not normally in more sparsely populated areas of the world

Status Population: several billion

Brown Rat

Rattus norvegicus

The brown rat is one of the most successful mammals on the planet. It rivals humans in terms of distribution and number, and continues to exploit us successfully despite our best efforts to exterminate it.

THE BROWN RAT IS PERHAPS the most reviled of all mammals. Even the word "rat" has all kinds of meanings, every one of them negative. Someone in a bad mood is described as "ratty," we detect wrongdoing when we "smell a rat," and telling tales to get someone else into trouble is "ratting."

Rats have few friends in spite of the fact that the vast majority of species are totally harmless. The brown rat is one the largest of 56 species in the genus *Rattus*, most of which live completely wild and never trouble people at all. But such is the strength of the brown rat's bad reputation worldwide that virtually all rats and ratlike rodents are treated as vermin.

Eastern Origins

Despite the misleading alternative name of Norway rat, the brown rat is thought to have originated in India or northern China. It spread to Europe and the Americas less quickly than its more inquisitive cousin, the ship rat, but today it is the dominant commensal rat species in most temperate parts of the world. In many places it has displaced the ship rat altogether. Brown rats prefer to live on the ground rather than in trees, and their liking for wet places suggests that their natural habitat may once have been stream banks. Brown rats swim well and are often associated with canals, sewers, and irrigation systems. They are expert at catching fish. They are also proficient diggers, and in the wild they create extensive burrow systems with many entrances and chambers.

The diet of brown rats can be extraordinarily diverse. They even manage to survive on the debris of seashores. Given a choice, however, they seem to prefer eating

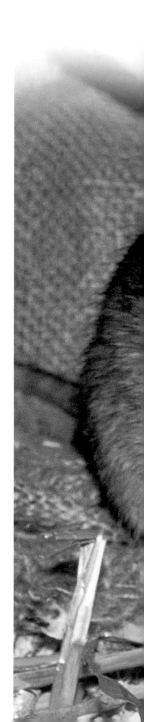

meat and animal matter to fruit and grains. Their teeth are relatively unspecialized: They use their incisors for gnawing and their molars for grinding fragments of food. In association with humans rats will eat almost anything, including soap, wax, leather, and paper. They thrive in cities, where drains provide them with shelter and food. Litter and overflowing trashcans offer all kinds of high-energy fatty foods loved by humans and rats alike. Food is often so abundant that an urban brown rat can spend its entire adult life in a home range as little as 80 to 500 feet (24 to 150 m) across.

The social structure of brown rats is rather variable, with the level of organization depending on the density of the local population. At low densities dominant males defend territories within which several females will collectively and cooperatively rear his offspring. If all available territories are taken, the excess rats form large gangs within which many males try to mate with every estrous female. Aggression is rife, there is no fixed dominance structure, and the stresses of life mean that reproductive success is low compared with that in the well-organized world of the territory-holding rats.

Rapid Breeders

Under good conditions brown rats can breed prolifically. A single pair can, in theory, multiply to over 15,000 animals in the space of a year. Rapid increases in population cannot be sustained for long, however, and brown rat numbers tend to fluctuate considerably. Overcrowding either leads to population crashes caused by starvation or disease, or to sudden mass dispersions, with thousands of rats suddenly on the move.

Brown rats will fight fiercely for their lives and will attack dogs, cats, and even people if cornered. There are even reports of brown rats killing babies and helpless humans by biting them continuously until they bleed to death. Of course, such incidents are extremely rare, but they add to the rat's evil reputation.

It surprises most people to discover that brown rats are actually quite clean animals. Wherever possible they take pains

⊕ Brown rats can feed on a wide variety of foods. They are a versatile species and can make their living almost anywhere there is human habitation.

to groom themselves using their tongue, teeth, and claws to wash and scrape dirt from their fur. They cover their palms in saliva to wipe their face and whiskers clean. However, such is the filth in some of the man-made environments they frequent, it is virtually impossible for some rats to ever get fully clean. It is true that rats carry many diseases that can infect people. Rabies, typhus, Weil's disease, rat-bite fever, and food poisoning (*Salmonella*) are just a few of the more serious infections known to be spread by brown rats.

Pest Control

The battle to control rats has been running for centuries. People have been employed as rat catchers since the Middle Ages, and there were dogs trained specifically for the task. The legend of the Pied Piper of Hamelin tells how the mysterious piper led the town rats to their deaths in the local river, having bewitched them with his music. When the townspeople refused to pay him, the piper took revenge by piping away all their children, who were never seen again. The rats in the story were probably ship rats rather than brown rats, but either way it normally takes more than music to rid a town of rat infestations. However, rats are vulnerable to poisoning. They cannot vomit; so even if they realize something is making them ill, they cannot void it from their stomach. One of the earliest rat poisons

⊕ *The fairytale of the Pied Piper of Hamelin tells of a piper who charmed the rats away from the German town with his mysterious music.*

⊝ *A brown rat carries an infant. The young are born blind and naked but are quick developers—after just three weeks they are ready to leave the nest.*

Even Rats Have Their Uses

Albino brown rats are naturally more docile than their full-color relatives. After generations of selective breeding they have become exceedingly tame and are widely used in medical and scientific research. Twenty million white rats are used in United States labs every year. Some are bred with specific weaknesses so that medical researchers can assess the efficacy of various new therapies. Others are used in experiments on physiology, neurology, genetics, behavior, and psychology. Brown rats have even been into space. In 1960 two lab rats spent time in orbit aboard a Russian satellite and returned to earth apparently none the worse for their adventure.

⊕ *An albino brown rat in a piece of laboratory equipment being used in a toxicology test.*

was derived from a plant called Mediterranean squill. Eaten by a human or a dog, it causes severe nausea and vomiting, but in rats it causes paralysis and death. However, rats are smart, and once one rat has been poisoned by something, others in the colony will avoid eating the same thing. Also, rats will not eat anything that has made them feel ill before, so slow-acting or cumulative poisons are no use. Brown rats are also naturally suspicious of anything new. Ship rats are less cautious, which is one reason why poisoning campaigns have been more effective with that species.

Death Sentence

A breakthrough in rodenticide technology came in 1950 with the development of a poison called warfarin. Warfarin contains dicoumarol, a small dose of which causes massive internal bleeding and death. Most importantly, the rats are unable to detect warfarin in foods and so do not learn to avoid it. Nor do they learn from others' mistakes, because death occurs some time after the poison is eaten. Yet as early as 1958 there were examples of rats that were apparently unharmed by dicoumarol, and the percentage of resistant rats is growing. The development of new poisons continues.

Common name Deer mouse (white-footed mouse)

Scientific name *Peromyscus maniculatus*

Family	Muridae
Order	Rodentia
Size	Length head/body: 2.5–4 in (7–10 cm); tail length: 2–5 in (5–12 cm)

Weight 0.4–1 oz (11–28 g)

Key features Russet-bodied mouse with white underside and legs; tail 2-tone and variable in length; head large with huge, sparsely furred, round ears; black eyes and long whiskers

Habits Mostly nocturnal; stores food for winter; does not enter full hibernation

Breeding Three to 4 litters of 1–9 young born in spring and summer in north of range (anytime in south) after gestation period of 22–30 days. Weaned at 3–4 weeks; females sexually mature at 6 weeks, males rarely breed before 6 months. May live up to 8 years in captivity, rarely more than 2 in the wild

Voice Squeaks and buzzing sounds; drums forefeet on ground when excited

Diet Omnivorous: seeds and grains, fruit, flowers, and other plant material; also insects and other invertebrates

Habitat Varied; includes scrubland, prairie, desert, alpine areas, boreal forest, and woodland

Distribution All of North America except tundra regions of Canada and southeastern U.S; also Mexico

Status Population: very abundant

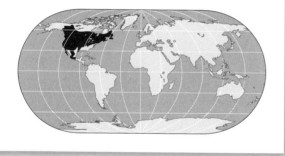

Deer Mouse *Peromyscus maniculatus*

The deer mouse is the most widespread member of its genus. It is among the world's more numerous mammals, producing up to four large litters a year.

THERE ARE OVER 50 SPECIES OF DEER mouse living in North and Central America, all of which belong to the genus *Peromyscus*. Many of them are similar and can only be told apart biochemically or by differences in the structure of their chromosomes. Others are geographically distinct. For example, a dozen or more are found only on small islands. At least one species of deer mouse has recently gone extinct, and several others are extremely rare.

Continuous Breeders

The common deer mouse, *Peromyscus maniculatus,* has no such survival worries. It occupies a huge range, including most of North America. In some years it is so abundant that it has been a severe nuisance to the forestry industry, nibbling seedlings and buds. Population booms usually follow so-called "mast years," during which trees such as oak produce a massive glut of nuts or acorns (mast). The excess food is stored by mice and many other animals and consumed over the following winter, boosting the survival rate. The mice may even continue to breed all winter long. They are in good condition in spring and can begin breeding early. Numbers rapidly build up as the progeny themselves begin to breed too.

Deer mice can reproduce extremely fast. A healthy wild female can rear four litters of up to nine babies a year, and her daughters can themselves breed before they are two months old. In captivity, where unlimited food and other luxuries make life easy for the mice, females can raise over 100 babies a year.

Every population boom is inevitably followed by a sharp decline. When a population crash happens, the mice spread out and become less social. Females in particular become partially territorial; and when food is scarce, they may try to kill each other's offspring. Feeding on young deer mice not only provides mother mice with a good meal, but it reduces the potential competition for her own family. Such rivalry between females is soon forgotten once winter rolls around again, and up to a dozen mice of all ages and sexes will nest together in order to share body warmth.

Wild deer mice are sociable creatures; and when the population booms, they will happily share home ranges. Mothers and daughters will even rear their young in the same nests. On such occasions the baby mice can end up being suckled by their mother, grandmother, aunt, or sister. Male deer mice often get involved in the care of their offspring too. They keep a careful eye on the young, retrieving any that wander off, and help keep the nest clean.

Deer mice are fastidious when it comes to matters of hygiene. Once a nest becomes soiled, they will leave it and build another. Nests are usually woven from grass and other plant material and lined with soft fibers such as thistledown. They are usually wedged into a secure spot such as a tree hole, disused burrow, or dense clump of vegetation.

Popular Laboratory Animals

The docile nature and clean habits of deer mice, along with their rapid rate of reproduction and nonspecialized diet, make them easy to keep in captivity. From a scientist's point of view they are ideal laboratory animals. The closely related white-footed mouse, *Peromyscus leucopus*, has been used as a model to investigate how male and female mammals compete to maximize their breeding success.

⊕ *Deer mice spend the day in burrows or trees. Occasionally, they come into buildings, where they nest in mattress stuffing or other soft material. On cold days they may enter a deep, torpid sleep, but they do not hibernate.*

Desert Wood Rat

Neotoma lepida

One of North America's most charismatic rodents, the timid desert wood rat has earned a place in folklore with its industrious building and sense of fair trade.

Common name
Desert wood rat (pack rat, trade rat)

Scientific name *Neotoma lepida*

Family Muridae

Order Rodentia

Size Length head/body: 6–9 in (15–23 cm); tail length: 0.5 in (1 cm)

Weight 0.4–1 lb (0.2–0.5 kg)

Key features Brownish-gray rat with pale underside and feet; ears round; furry tail up to half length of body

Habits Nocturnal, solitary, and timid; constructs large "houses" above ground, collecting building material from wide area; does not hibernate

Breeding Two or 3 litters of 1–5 young born at any time of year after gestation period of 30–40 days. Weaned at 4 weeks; sexually mature at 2 months. May live up to 7 years in captivity, several years in the wild

Voice Generally silent

Diet Leaves, seeds, roots, and fleshy cactus pads; also insects and other small invertebrates

Habitat Desert

Distribution Deserts of southwestern U.S. and northwestern Mexico, including Baja California

Status Population: abundant

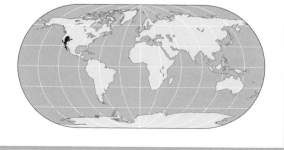

THE WOOD RAT'S ALTERNATIVE NAME of pack or trade rat refers to the species' most intriguing behavior. Like most rodents, wood rats are inquisitive and opportunistic. They will help themselves to almost any item of food or other useful object they come across and are always on the lookout for new pieces of building material to add to their elaborate nests.

Honest Traders

Bits of twig, cactus spines, and small bones are all useful nest-building materials. They are collected from all over the rat's home range, an area of about 4,400 square feet (400 sq. m). Wood rats are also very fond of bright, shiny objects, such as foil, silver cutlery, and glass, and will requisition such objects from campsites and gardens. Their habit of collecting objects is not unusual, since rodents are well-known thieves. Where the wood rat differs is that it often appears to "pay" for what it takes by leaving some other object in its place. What actually happens is that the wood rat collects one item and is in the process of carrying it home when it finds something better. It cannot carry both objects, so it drops its original prize in favor of the more desirable one.

Wood rats are careful and discerning architects. Their nests can be huge, up to 5 feet (1.5 m) across and the same in height. They are often built around the base of a spiny cactus. The spines are built into the fabric of the nest so that it is difficult for another animal to get inside. Desert wood rats are experts at moving among the spines and can do so very quickly without ever seeming to injure themselves. The nest offers a refuge from the extremes of heat and cold typical of the desert climate. It is also impregnable to most predators. The nest may contain several chambers. Some are lined with soft material and used for sleeping, while others serve as larders. A wood rat can build a basic nest in about a week, but often it does not have to start from scratch. Many nests are used time and again by generations of wood rats, each extending or modifying them to their own personal specifications. Some nests are known to be hundreds of years old. Desert wood rats eat all kinds of plant material, but succulent cacti are especially important because they provide moisture as well as food. Desert wood rats rarely get the opportunity to drink free water.

Life is particularly hard for nursing mothers who must also find enough water to convert into milk for their young. The mothers of large litters sometimes die in the effort to sustain their offspring. But as a rule, wood rat litters are fairly small, and females generally produce only two or three litters a year. The young disperse once they are weaned.

A desert wood rat in the southwestern United States. The alternative name of "pack rat" refers to the animal's habit of transporting objects around its range.

Endearing Pets

Adult wood rats live solitary lives, although they are nonaggressive and nonterritorial, and their home ranges overlap with those of their neighbors. They can be tamed if captured when young and make endearing pets. Nonetheless, they can also create a nuisance to farmers and homeowners who do not wish to trade their electrical wiring, nuts and bolts, and even silver jewelry for gifts of twig and bone. However, the animals are rarely numerous enough to be considered a serious pest.

Common name
Southern red-backed vole

Scientific name *Clethrionomys gapperi*

Family	Muridae
Order	Rodentia
Size	Length head/body: 3–5 in (8–13 cm); tail length: 1–2 in (3–5 cm)

Weight 0.2–1.5 oz (6–43 g)

Key features Rounded body with short legs; tail about half length of body; fur grayish-brown with reddish flush along back; face short with prominent black eyes and large, rounded ears

Habits Active day and night and all year round; territorial; climbs well

Breeding Up to 4 litters of 1–11 young (litters larger in north) born early spring to fall after gestation period of 17–20 days. Weaned at 3 weeks; sexually mature at 3 months. May live over 4 years in captivity, up to 20 months in the wild

Voice Soft, chirping alarm call and audible clattering of teeth

Diet Fruit, nuts, fungi, lichen, green shoots, and other vegetation; also insects

Habitat Tundra, moorland, mossy forest undergrowth, and woodland floors

Distribution Southern Canada from British Columbia to Newfoundland; northern coterminous U.S. south to Arizona

Status Population: abundant

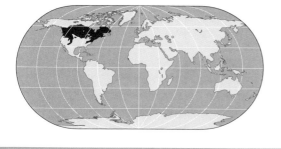

Southern Red-Backed Vole
Clethrionomys gapperi

The southern red-backed vole is typical of the widespread genus Clethrionomys. *Similar voles occupy much the same kind of habitat in other parts of North America and in Europe and Asia.*

AS A GENERAL RULE, MEMBERS OF the vole branch of the mouse family—which includes the lemmings—are creatures of high northern latitudes. Their chubby bodies, short legs and tail, and small, neat ears are adaptations to a cool climate, since they reduce the surface area through which body warmth can escape.

Chestnut-Red Stripe

Red-backed voles are ideally suited to life in cool, damp climates such as the Canadian tundra and boreal forest. Farther south suitable habitat is restricted to the mountains of the Rockies and Appalachians. The southern red-backed vole characteristically has a thick coat of muted gray-brown fur, typically highlighted with a broad stripe of rich chestnut-red running from head to tail down the animal's back.

Southern red-backed voles are opportunist feeders, and their diet tends to reflect seasonal availability. In spring they eat mostly green shoots and tender new leaves, while in summer energy-rich fruits take precedence. Later they take advantage of seeds and nuts, storing any they cannot eat to help see them through the lean times ahead. The vole's teeth, including its grinding molars, continue to grow well into adult life. When the molars eventually stop growing and form roots, the vole's days are numbered, since the teeth become worn out by munching on tough vegetable matter. Once they stop working properly, the vole will starve. But few voles survive long enough for their teeth to let them down. Most die of other causes at an earlier age. The voles are preyed on by a wide variety of mammals and birds.

Except for females with young, red-backed voles live alone in well-spaced home ranges, which vary in size depending on sex, habitat quality, and season. Winter territories are small, as little as half an acre (0.2 ha), because snow restricts the voles' movement. In summer the voles may occupy ranges 10 times as large. It is a considerable area for such a small animal, but the voles are quite fiercely territorial. They will drive out intruders, including animals of other species. Scent markers play an important part in staking out territories, as does sound.

The voles utter sharp chirping calls to advertise their presence to other voles. If the resident vole is killed, its territory will not remain vacant for long. When moving around their territory, red-backed voles tend to bound along open routes. They use fallen logs or runways created by the regular movements of other larger animals, but they do not stray far from cover. Red-backed voles remain active during the winter, creating personal networks of tunnels under the snow from which they can forage for fungi, lichen, and tree bark.

The reddish coat distinguishes this group of species from other voles. Unlike mice, typical voles have a snub-nosed appearance and ears that do not project far out of the fur.

Starting Early

Young red-backed voles are able to breed at the age of three months, but young born late in the summer may have to wait until the following year before making their first attempt. The breeding season starts surprisingly early, often in late winter, when there is still a covering of snow on the ground. However, if a female is in good condition, there is no reason why she cannot rear a healthy litter, which will be ready to emerge as soon as the snow melts away. The young of that litter will themselves be breeding within a few weeks.

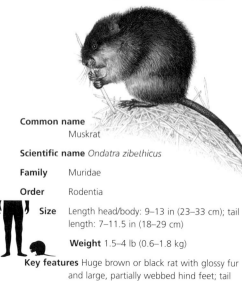

Common name
Muskrat

Scientific name *Ondatra zibethicus*

Family Muridae

Order Rodentia

Size Length head/body: 9–13 in (23–33 cm); tail length: 7–11.5 in (18–29 cm)

Weight 1.5–4 lb (0.6–1.8 kg)

Key features Huge brown or black rat with glossy fur and large, partially webbed hind feet; tail naked and slightly flattened from side to side

Habits Mostly nocturnal and crepuscular; social but bad tempered; semiaquatic

Breeding Up to 6 litters of 1–11 young born at any time of year (spring and summer in north of range) after gestation period of 25–30 days. Weaned at 2–3 weeks; sexually mature at 6–8 weeks in south of range, later in north. May live up to 10 years in captivity, rarely more than 3 in the wild

Voice Growls when annoyed

Diet Mostly aquatic plants such as reeds, rushes, cattails, and water lilies; also grasses and animal matter, including fish and shellfish

Habitat Pools, lakes, rivers, marshes, and swamps with plenty of plant life

Distribution Southern Canada and most of U.S.; patchy in California and Texas; introduced in Eurasia and South America; introduced British population exterminated within a few years

Status Population: abundant—many thousands, perhaps millions. Trapped for fur and meat; considered a pest in parts of introduced range in Europe

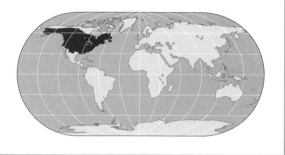

Muskrat

Ondatra zibethicus

The muskrat is not really a rat at all, but a kind of large vole. Its native home is North America, but thanks to a little help from humans it now also occupies huge areas of mainland Europe and Asia.

COMMON MUSKRATS ARE LARGE rodents that feed mostly on water plants, which they gather beneath the surface of still- and slow-flowing rivers and lakes. Anatomical clues to their semiaquatic lifestyle include a tail that is flattened from side to side for use as a rudder and partially webbed hind feet. The feet also bear a fringe of stiff hairs on the outer edges, which act a bit like fins. The hairs increase the surface area the muskrat can use to paddle through the water. They may also help grip slippery floating logs and perhaps stop the animal from sinking in soft mud. Muskrats are superb swimmers and divers. If need be, they can remain submerged for over a quarter of an hour, although most dives are much shorter— about three minutes on average.

Muskrat Architecture

Depending on their habitat, muskrats live either in waterside burrows or in houses built of plant stems and mud. Burrows are dug directly into the banks of rivers. They are difficult to see because the entrances are well below the water level. The reason for positioning the entrances so far down is to ensure that they are not exposed during low water or frozen over in the winter. Shallow water that freezes from top to bottom in winter is not suitable for muskrats.

Where there is no bank, in areas of marshland, for example, the muskrats build one of their own by collecting large quantities of plant material into a heap and covering it with a thick layer of mud. The mound rots down into a dense mass within which the muskrat creates a nesting chamber above the water line. Like burrows in a natural bank, the "houses" are accessible only from below.

In the far north of the muskrat's range, where pools and rivers are frozen over for much of the winter, the muskrat also builds temporary shelters over holes in the ice. These "tents" provide the muskrat with a secluded spot in which to feed and rest between foraging dives.

Muskrats live alone or in family groups, members of which help defend territory and build and maintain the houses if necessary. While suckling a litter of babies, the female excludes her mate from the breeding nest, forcing him to live in a nearby burrow or a separate nest attached to the main house. Muskrats are potentially prolific breeders, but litter sizes vary geographically. In the south litters are small, but there can be as many as six in a season. In the north families are larger, but there are fewer litters per year.

In the last hundred years muskrats have been intensively hunted and trapped for their glossy, luxurient fur. At its peak in the early 1970s the North American fur trade marketed over 10 million muskrat skins a year, worth more than $40 million. Demand for fur has since declined, but the muskrat is still an economically important animal.

Widespread Aliens

It was their status as fur-bearers that encouraged humans to introduce muskrats to several places outside their natural range, and the species is now a widespread alien in Europe, Asia, and South America. Muskrats were introduced to fur farms in Britain between 1927 and 1932. Many escaped and soon spread widely. Over 4,500 were trapped, and the species was eliminated by 1937. Elsewhere, some populations now support a profitable fur trade, while others have caused significant damage to wetland ecosystems. Muskrats are especially unpopular in the Netherlands. Here their burrowing and feeding activities threaten to destabilize the dikes that protect large areas of reclaimed agricultural land from flooding.

⊕ Muskrats are well adapted to a life in and around water. They have partially webbed hind feet and a flattened tail, which they use as a rudder. Here muskrats feed on aquatic plants.

North American Porcupine

Erethizon dorsatum

The North American porcupine's bumbling movements, loud snufflings, and tousled coat are an endearing combination. Yet its comical appearance is misleading, since far from being slow-witted and clumsy, the animal is in fact rather smart.

THE PORCUPINE'S QUILLS ARE actually enlarged hairs, thicker and stiffer than normal hairs, but made of the same substance (a protein called keratin) and growing from the same sort of cells. In Old World porcupines the quills grow in clusters, but those of the North American porcupine grow singly among the normal hairs. The longest quills sprout from the porcupine's rump; short ones grow almost everywhere else except on the belly, including on the animal's head and cheeks. These smaller quills are usually hidden by the porcupine's fur, and especially by the long, pale guard hairs that give the animal its distinctive spiky hairdo.

Porcupines have very acute senses of smell and hearing, and will pause frequently in their activities to sniff the air before moving slowly on. They have a large brain and a good memory, and they can be trained quite easily.

Not Invincible

Despite their impressive defenses, porcupines have predators, notably large mustelids (fishers and wolverines), coyotes, bobcats, and large birds of prey. Successful porcupine hunters have learned to flip the animal over and attack it from the nonspiny underside. In parts of North America where porcupines create a nuisance to farmers and plantation owners, fishers have been used to help control their numbers.

Female porcupines are generally less sociable than males and behave aggressively

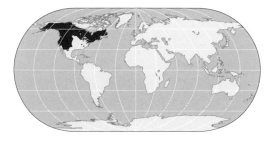

Common name North American porcupine (quillpig)

Scientific name *Erethizon dorsatum*

Family Erethizontidae

Order Rodentia

Size Length head/body: 25–34 in (64.5–86 cm); tail length: 5.5–12 in (14–30 cm)

Weight 7.7–40 lb (3.5–18 kg)

Key features Large rodent covered in stiff, brown to black fur; long, yellow spines on head, back, rump, flanks, and tail; feet have long, thick claws; short face with small, dark eyes and small ears hidden in hair

Habits Generally nocturnal; partially arboreal; climbs carefully; swims well; lives alone or in small groups when breeding or sheltering in winter

Breeding Single young (rarely twins) born April–June after gestation period of 7.5 months. Weaned at 2–6 weeks; sexually mature at 2 years. May live about 18 years in captivity, similar in the wild

Voice Grunts, growls, coughs, barks, and whines; also makes clattering sound with teeth

Diet All kinds of plant material, including leaves and shoots, buds, flowers, fruit, seeds, nuts, twigs, bark, and wood; also gnaws bones

Habitat Mixed forest; also tundra, farmland, scrubland, and desert close to wooded areas

Distribution North America from Alaska throughout most of Canada and the continental U.S. south to northern Mexico and the Carolinas

Status Population: abundant

Lethal Weapons

A commonly held belief about porcupines is that they can shoot their quills directly into an enemy, but that is not true. The only way a porcupine can spike another animal is by coming into direct contact. When threatened, a porcupine will turn its back and raise the quills on its rump. It scuffs its feet and thrashes its short tail at its attacker. The quills are easily detached from the porcupine's skin, and their barbed tips are so effective at snagging the flesh that one quick flick of the tail can leave several quills deeply embedded. The barbs allow the quill to slip easily into muscle, and over the course of a few hours a quill can work its way deeply into the body, sometimes even penetrating vital organs, with fatal results.

toward other porcupines. Female home ranges are exclusive, but those of males overlap with each other and with the territories of females. For most of the year males interact peacefully and stay out of the way of the females. Dominance among males is established by display and size, rarely by fighting. During the fall, when females come into estrus for less than a day at a time, the dominant local male will approach and begin an elaborate courtship ritual. The male and female dance around each other making a variety of calls, and the male culminates the display by spraying the female with urine. This usually impresses her enough to allow him to mate, but afterward he is driven away almost immediately.

Care and Attention

Porcupines are what biologists call "k strategists." This means they breed slowly, lavishing care on usually just a single offspring at a time (twins are rare). Pregnancy lasts about seven months; newborn porcupines are large—they can weigh up to 10 percent of the

⊕ *A North American porcupine feeds on pine needles. The animals can do serious damage in plantation forests, especially in the winter.*

mother's own body weight. Relatively speaking, this is about twice as big as an average human baby. They are also remarkably precocious, being born fully furred, with their eyes open and able to walk immediately. Within a day or two they can even climb, and they instinctively know how to defend themselves, turning their back and raising their quills even before they have had a chance to harden.

Weaning starts early, with most young porcupines beginning to eat leaves practically immediately. After just two weeks the youngster can survive without its mother's milk. Nevertheless, it is usually allowed to nurse for at least another month. It takes the young porcupine three or four years to reach full size, although sexual maturity comes a little sooner. The slow reproduction rate in porcupines is compensated for by the fact that the survival rate of infants is high. Porcupines are also very long-lived rodents. Most individuals live well over 10 years, and some as long as 18.

Tree Foragers

One of the most important lessons for any young mammal to learn is what to eat. Fortunately for the porcupine, this seems to be almost any kind of plant. The vast majority of the porcupine's diet is vegetarian, and much of its food is found in the trees.

Porcupines climb extremely well, but their movements are slow and carefully controlled, very different than the rapid, scrambling style of squirrels. The porcupine does not leap from branch to branch or dash headlong up and down treetrunks. Instead, it clutches the trunk with all four muscular legs, its long claws and naked soles and palms providing good grip. It hauls itself up in a manner not unlike that of a koala. Once in the branches, it walks slowly and carefully, maintaining excellent balance despite the absence of a long tail. Porcupines can also swim well—their hollow quills make them very

⊕ *Despite its clumsy appearance, the North American porcupine is an excellent climber and frequently ascends great heights in search of food such as berries and nuts.*

Look After Your Teeth

Cast-off deer antlers or bones scavenged from the carcasses of dead animals are often carried back to the porcupine's den, where they can be gnawed at leisure. The behavior is beneficial in two ways—it provides a valuable supplement of calcium and other minerals, and helps keep the porcupine's front teeth in good shape. Like all rodents, the North American porcupine has a single pair of chisel-shaped incisors in each jaw, the front surfaces of which are coated with hard enamel. Behind the enamel the rest of the tooth is made mostly of dentine, a slightly softer material that wears away faster than the enamel, leaving a sharp edge at the front of the tooth. All four incisors continue to grow throughout the porcupine's life, so a steady rate of wear is important in keeping the teeth at the optimum length. In winter, when the porcupine's diet includes a lot of wood, the incisors are worn away steadily, but summer foods are a lot easier on the teeth. Fruits and nuts have much greater nutritional value than wood and bones, but the porcupine continues to gnaw these hard materials as part of its dental care routine.

buoyant, so all they have to do is paddle with their feet to make good progress.

Wide Wanderings

There are marked seasonal shifts in the North American porcupine's diet and foraging behavior. In summer it wanders widely, sometimes using a home range of more than 150 acres (60 ha). It feeds on as much energy-rich food as it can find, including fruit, seeds, bulbs, and tubers. Males may wander even farther in the fall in search of estrous females. In winter, when long feeding expeditions would be risky and often pointless, they stay much closer to home and make do with pine needles and soft woody tissues stripped from beneath the bark of trees, especially conifers. Trees are easy to find in most porcupine habitat, and some animals never travel more than 33 feet (10 m) from their den all winter long.

Porcupines can cause serious damage to plantation forests, especially in winter, when they turn their attention to softwood trees.

⊕ *A porcupine sits on snow-covered ground. When the winter arrives, porcupines rarely wander far from their den to forage, preferring to make do with pine needles and tree sap.*

They start gnawing off the bark at snow level, working their way up until they have exposed the wood all around the tree. What they are after is not the bark itself but the soft, sappy tissues underneath. When those have been removed, the flow of water and nutrients to the top of the tree is cut off, and it will die.

It is not known exactly how much porcupines cost the North American forestry industry each year, but they are not by any means the only culprits—rabbits, deer, and squirrels also strip bark. In addition, porcupines can create a nuisance around people's homes, usually by gnawing wooden fences and furniture or other objects such as tools. They are especially attracted to salty things and appear to take great delight in chewing up sweaty saddles or leather gloves!

⊜ A porcupine in a field of spring flowers. In the spring the animals frequently venture onto meadows to feed on grasses during the evening hours.

Porcupines in Folklore and Legend

Porcupines feature in all kinds of traditional stories, some of which are based on truth; others are pure fantasy. Native American folklore tells how the porcupine used its cunning to kill a buffalo. The porcupine persuaded the buffalo to carry him across a river in its belly. Once across, the porcupine killed the buffalo by spiking its heart with his quills. Having emerged from the dead buffalo the porcupine tricked a coyote into carving up the meat for him. The coyote was angry when he saw the porcupine eating the meat, so he killed him. But the porcupine came back to life and escaped up a tree with the meat. Then he killed the coyote and his family, all except one young cub. The porcupine and the cub became friends and worked together to hunt buffalo in the future.

The cunning and bloodthirsty animal in the story seems a far cry from the placid vegetarian porcupine, but a few elements of the story ring true: Porcupines are smart animals; their quills are sharp enough to kill; they sometimes play dead when threatened; and they appear to come back to life when the danger has passed. Last, they are excellent climbers and will seek refuge in trees. The association between coyotes and porcupines is not normally friendly, and coyotes are one of the few American carnivores that sometimes kill and eat porcupines. However, where plenty of other prey is available for coyotes, they may view porcupines as more trouble than they are worth and leave them alone.

The 10-Year Cycle

Populations of snowshoe hares fluctuate wildly. Highs and lows seem to occur roughly every eight to 11 years. The phenomenon was first recognized by fur trappers who named it the "10-year cycle" and puzzled at its cause. For a long time it was thought that the hares were periodically wiped out by some mysterious disease, and a few people still refuse to eat hare meat for fear of infection. In peak years the population density can exceed 7,500 hares per square mile (3,000 per sq. km), as happened in Alberta in 1970. These massive highs are inevitably followed by dramatic declines, and five years later the Alberta hares numbered a mere 6.25 per square mile (2.5 per sq. km). The population falls when overcrowding creates food shortages, especially in winter. Adult females respond by delaying the start of their breeding season and raising fewer, smaller litters. The decreased birthrate no longer keeps up with the losses due to predation. The ratio of predators to prey soars, and the hare population plummets. Soon there are so few hares left that predators begin to starve, and their numbers crash. Meanwhile, vegetation eaten by the hares has had a chance to recover, and with predator numbers very low, another boom time for hares ensues.

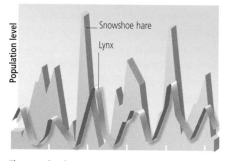

The snowshoe hare makes up 80 to 90 percent of the Canadian lynx's diet. When the hare population drops, lynx numbers decline soon after.

relish green grass and succulent fresh herbage such as vetch, wild strawberry, fireweed, and clover. In winter they rely heavily on conifers including larch, cedar, and various pines, and broadleaves such as maple, aspen, birch, willow, and hazel. They nibble young saplings and new twigs, bite off leaf buds, and strip the bark of older trees. In very harsh winters they have been known to eat carrion, including the remains of other hares that have died of cold or starvation. As in all other lagomorphs, complete digestion requires food to make two journeys through the hare's gut, resulting in two different kinds of droppings. The first kind are soft, moist pellets, which are immediately eaten again. The second pellets are dry and hard and contain only indigestible waste material.

⊙ *A snowshoe hare in the process of molting. Each winter snowshoe hares molt into a white coat for camouflage. Hares living outside areas with continuous snow generally do not molt.*

Rain Haters

Snowshoe hares are active all year round, but avoid going out in wet and windy weather unless absolutely necessary. When food is in short supply, such as in late winter or during periods of high population density, they may be forced to wander several miles a night in order to find enough to eat.

However, adult hares are generally rather sedentary animals and will spend their whole lives in the same familiar area. Snowshoe hares

⊙ *A lynx carries off a snowshoe hare. When hare numbers are low, the shortage of food leads to a fall in the number of lynx kittens born.*

use a home range of about 14 to 30 acres (6 to 12 ha), with little difference between the sexes. Unlike some other species of hares, snowshoes seem not to like open spaces and rarely venture far from cover, except in population boom years when animals are forced to spread out. Dispersing animals are nearly always juveniles, driven to seek new homes when food becomes short in their mother's range during their first winter. Dispersal journeys rarely exceed 5 miles (8 km), but are nevertheless fraught with danger. On average about 80 percent of young snowshoe hares never live to see their first birthday, and in some years the juvenile death rate may be as high as 97 percent.

Snowshoe hares are solitary in most of their activities, but they are tolerant of others living in overlapping home ranges. They will share feeding sites and use the same forms for resting, although they do so at different times. Males tend to occupy larger ranges than females, anything up to 25 acres (10 ha), and the males in a given area will fight to establish and maintain position in a simple linear hierarchy. Rival males fight by boxing and kicking with their hind legs, and such contests occur with increasing frequency as the breeding season approaches. During the breeding season

Snow Shoes

Getting around in deep snow can be tricky, which is why people living in high northern latitudes invented snowshoes. The shoes work by effectively enlarging a person's feet, so spreading their body weight over a wide area. The wearer can now move over the surface of soft snow without sinking. Many animals living in snowy places use a similar tactic—for example, polar bears, snow leopards, and lynx all have relatively large feet. The snowshoe hare's feet are not particularly broad, but the long toes are flexible and can be spread wide to create a "snowshoe" when required. The feet are also very furry, which helps spread the animal's weight even further and also prevents frostbite. With its built-in snowshoes a snowshoe hare can scud away over the surface of the snow, leaving small-footed predators such as coyotes and foxes floundering in its wake.

several males will gather around a female, but the highest-ranking hare will usually chase off the competition and stay close to the female until she is ready to mate.

Synchronized Breeding

The precise timing of the breeding season varies from year to year and place to place, but is highly synchronized between females living in the same area. All females come into season within a few days of each other, so the first litters all arrive about the same time. That provides a degree of protection from predation.

Provided the hare population is relatively large, the simultaneous arrival of so many young provides more prey than the local predators can eat, so some hares will always escape death. Because female snowshoe hares usually mate again within hours of giving birth, subsequent litters also arrive simultaneously throughout the season.

Home Alone

As the time approaches for a female to give birth, she becomes more aggressive, especially toward males. The litter is born at a secluded but open spot, sometimes in a depression of flattened vegetation.

Newborn hares have a full coat of fur and open eyes. They are able to move around by themselves and creep into dense vegetation to hide alone during the day. In the evening they congregate at a meeting place where their mother suckles them for five to 10 minutes. The young grow fast, increasing their birth weight by five or six times before they are weaned. They are independent at three weeks and reach full size after five months. However, they do not usually breed until the following year.

⊕ *A running snowshoe hare shows off the large size of the animal's feet. Oversized feet are an adaptation to living in deep snow and help the hare outrun predators.*

Common name Black-tailed jackrabbit

Scientific name *Lepus californicus*

Family	Leporidae
Order	Lagomorpha

Size Length head/body: 18.5–25 in (47–63 cm); tail length: up to 4 in (10 cm)

Weight 3.3–4.4 lb (1.5–4 g)

Key features Large hare with huge, erect black-tipped ears; legs long and slender; face has flat profile and large bulging eyes; fur grizzled gray-brown, paler below; dark dorsal stripe blends into black tail

Habits Active mainly in the evening; social; terrestrial, fast-moving, and nimble

Breeding Two to 6 litters of 1–6 (usually 3 or 4) young born at any time of year after gestation period of 41–47 days. Weaned at about 3 weeks; sexually mature at 7–8 months, but will often not breed until following year. May live up to 6 years in captivity, 5 in the wild

Voice Generally silent

Diet Grass and other plants of arid zones, including sagebrush, cactus, juniper, and mesquite; may raid cereal crops and orchards

Habitat Desert, prairie, and pasture in arid and semiarid areas

Distribution Southwestern United States and Mexico

Status Population: abundant, but fluctuates on 10-year cycle. Has increased in range and population since European settlement thanks to predator control

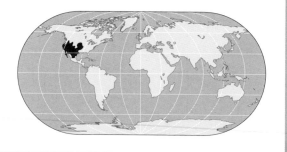

Black-Tailed Jackrabbit

Lepus californicus

North America's largest hare was named the "jackass rabbit" by early settlers on account of its enormous ears. The name was shortened to jackrabbit and is now also applied to many related species.

BLACK-TAILED JACKRABBITS are adapted to life in the hot, dry deserts of the American West. During prolonged drought the animals can eke out a living on meager rations of creosote and mesquite scrub, gaining all the water they need to survive from the succulent leaves of cacti. The jackrabbit is one of a few native wild mammals to have benefited from the spread of European settlers throughout North America. The development of land for agriculture has provided it with more food and also removed many of its predators.

Cooling System

Jackrabbits spend the hot part of the day sheltering in thickets or in the cool burrows of other animals. While they generally do not dig themselves, they are known to occasionally create burrows to avoid extreme temperatures. The Cape hare does the same to escape high desert temperatures. Jackrabbits emerge at dusk when there is less danger of overheating. The huge ears act as radiators, dispersing body heat to keep the animal cool. Large veins in the thin tissues of the ear allow the blood to be cooled in the breeze as the hare moves around.

Jackrabbits are social in that they live in overlapping home ranges and recognize each other as individuals. Breeding rights are largely determined by status, which is achieved by fighting. However, there is no close-knit family life, and individuals do not seek out each other's company. Young jackrabbits are born very well developed. Apart from a brief bout of frenzied suckling once a day for their first three weeks of life, they have little contact with each other or their mother. Adult jackrabbits gathering in one place usually only do so to exploit the same feeding resource. Large numbers sometimes build up on sparsely vegetated pastures, where they compete with cattle and sheep for limited food. Such groups intensify the problem of overgrazing and make the jackrabbit unpopular with ranchers. Numbers were controlled by having large drives, where the jackrabbits would be herded into fenced areas and shot. In this way up to 6,000 might be killed in a single day.

Black-tailed jackrabbits are highly susceptible to infections caused by the bacterium *Pasteurella tularense*, which also affects humans. The bacterium results in a nasty disease known as tularemia—after California's Tulare County, where it was first described. In some years over 90 percent of the jackrabbit population is infected, and the majority will die. In humans the disease can be treated with antibiotics, but nevertheless it provides another reason for jackrabbits to be controlled.

Losing Battle

Controlling jackrabbits is difficult. Organized culls in which the animals are shot or netted and killed can reduce local numbers. But since jackrabbit populations are naturally unstable, the species is accustomed to bouncing back from low levels. Low population densities merely encourage the remaining animals to breed faster, and farmers often find they are fighting a losing battle.

Despite being regarded as a nuisance in parts of its natural range, the black-tailed jackrabbit has been introduced to other parts of the United States as a game animal. There are now introduced jackrabbits living on Nantucket Island off Massachusetts as well as in parts of New Jersey and Kentucky.

⊕ *A black-tailed jackrabbit showing the distinctive black patch above its tail. The huge ears—another distinguishing feature—help keep the animal cool.*

Common name Arctic hare (polar hare, Greenland hare)

Scientific name Lepus arcticus

Family	Leporidae
Order	Lagomorpha
Size	Length head/body: 20–21 in (51–53 cm); tail length: 1.5–4 in (4.5–10 cm)

Weight 5.5–15 lb (2.5–7 kg)

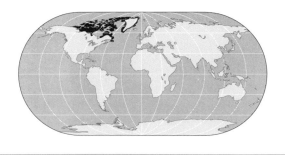

Key features Gray or brown in summer with white tail; white all over in winter, except for tips of ears, which are black

Habits	Usually solitary, but sometimes forms large groups; normally lives out in the open or among boulders
Breeding	One litter of 2–8 young (average 5) born June or July after gestation period of 53 days. Weaned at about 8 or 9 weeks; sexually mature by following summer. May live about 7 years in the wild, not normally kept in captivity
Voice	Normally silent
Diet	Woody plants, moss, lichens, roots and flowers of willows; also some grasses
Habitat	Tundra
Distribution	Far north of Canada and Greenland south to Hudson Bay and Newfoundland
Status	Population: total figure unknown. Common

Arctic Hare

Lepus arcticus

Throughout its range the Arctic hare is the only hare present. It lives in one of the world's most hostile environments, but is well adapted to survive there.

THE ARCTIC HARE IS WIDELY distributed in the tundra regions of northern Canada and Greenland from sea level to about 3,000 feet (900 m). In Canada the Arctic hare is common on Ellesmere Island and widespread south to the rocky plateau of eastern Newfoundland. In western Greenland it occurs in most of the ice-free coastal areas, but is sometimes found on the ice itself, more than a mile away from land.

Cold-Weather Adaptations

Like other mammals of the far north, the Arctic hare is well adapted to survive in the cold and barren conditions that characterize such places. It has relatively short ears for a hare and therefore a smaller surface area through which to lose heat. In addition, the blood circulation is arranged in such a way to minimize the amount of warm blood that goes into the ears. Arctic hares crouch down in a compact shape to conserve heat and try to avoid the chilling effect of the wind by sheltering as much as possible. The feet are large and hairy, helping prevent the animal from sinking into soft snow. The hare often lives beyond the tree line, although in winter some individuals retreat to the edge of the northern forests.

The Arctic hare uses its incisor teeth to nibble at the lichens and mosses growing among rocks. However, it feeds mainly on woody plants, taking buds, berries, bark, and roots. Willows provide the favorite food and form almost the entire diet during winter, while in summer they are supplemented with grass. Arctic hares are likely to be significant competitors of muskox and reindeer (caribou) during the winter when all three species face a shortage of food. In the winter months Arctic hares will brush away shallow snow using their

paws to uncover buried grass and lichens. When the snow has a thick frozen crust, the hare will use its hind feet to break through, pushing pieces aside with its nose.

Mating usually occurs in April and May, and the young are born fully furred and with their eyes open seven to eight weeks later. They are protected from the elements by a nest made of dry grass and moss, often lined with fur plucked from the mother's own belly. The nest is usually among boulders, but may be in a short tunnel. Females have only a single family each year, generally with about five young. The offspring have gray fur that helps camouflage them among the rocks.

Following the birth of her babies, the mother will remain with them for two or three days continuously before going out to feed. The babies grow their first coat of white fur when they are about three weeks old, at which time they begin to become independent of their mother. She will continue to nurse them every 20 hours or so, spending up to four minutes feeding each one. The babies are fully weaned when they are about two months old, but face many dangers. Young hares are eaten by wolves,

⊕ An Arctic hare feeds in the tundras of the Arctic. Hares living in the most northerly regions may remain white throughout the year.

Being White

We are so used to animals being colored that it is startling to see a completely white one. Some animals are naturally white, like the Arctic hare. Stoats and Arctic foxes also turn white in winter, as do the willow grouse and ptarmigan. Such animals benefit from the camouflage and insulation properties of a warm, white winter coat in their snowy home.

Whiteness also helps hide the animal in wide-open spaces, where anything dark would show up boldly against the bright snow. Another useful feature of white is that it tends to radiate less heat than darker colors. As a result, the hare's winter coat retains valuable body warmth. The insulating layer of air trapped against the skin by the long, fine hairs provides extra protection from the cold. Whiteness is caused by minute air bubbles in hairs and feathers. Their presence scatters light in such a way as to appear white to our eyes. Normally pigments would obscure the effect, but in white animals the pigments are simply not there. There is no such thing as a white pigment. So an animal is white for the same reason as the sparkling laundry in television detergent advertisements: It reflects white light, having nothing to mask its basic whiteness.

Many other animals are white as a result of a freak event rather than regularly as in the Arctic hare. Such lack of color results from failure of the complicated biochemical processes that produce pigments. Sometimes the absence of color is brought about by inherited genes. The animal—called an albino—ends up wholly or partially white. True albinos do not form any dark pigment at all. They are therefore white all over and have pink or red eyes. Since there is no melanin formed in the retina of the eye, the pink blood vessels inside it are clearly visible through the pupil. Albino individuals have been recorded from a wide variety of species. They are often produced as a direct result of inbreeding in captivity.

Whiteness can also result from injury to the skin. This is the principle of freeze branding, in which a cold metal branding iron is used (instead of a red-hot one) to damage a patch of skin. The hairs that then regrow are white, since the normal pigment production mechanism has been knocked out in that area.

An Arctic hare rests on a patch of snow in the Northwestern Territories of Canada. The hairs of the winter fur are entirely white, although the animal still retains the black tips to its ears. Not surprisingly, the long, silky winter coat is highly prized in the fur industry, and large numbers of animals are trapped for their skins. However, Arctic hare meat is not considered to be especially palatable.

Arctic foxes, snowy owls, and birds of prey. The adults, on the other hand, are sufficiently fast and nimble that few predators can catch them.

Hare Locomotion

Arctic hares normally live alone, but they may gather in groups of more than 100 individuals, especially where there is plenty of food. The hares rarely venture far from rocks or other places that offer shelter from the wind or predators. They are fond of sitting and dozing quietly in sheltered spots warmed by the sun. Often two or more rest together. Arctic hares seem to like facing up a slope. If they run downhill, they turn around and face upslope again before resting. They move in a series of four-legged hops, each covering about 4 feet (1.2 m). Periodically, an animal will sit upright to survey its surroundings before moving on. Sometimes it will stand tall on its hind legs to get a better view. If necessary, an Arctic hare can run at up to 40 miles an hour (65 km/h) and sometimes bounds like a kangaroo with its forefeet barely touching the ground. It is also capable of swimming short distances to cross small rivers and streams. Male and female Arctic hares are similar in size and cannot easily be told apart from a distance.

⊕ *An Arctic hare kit hides in tall grass in the Yukon Delta National Wildlife Refuge in Alaska.*

Color Change

In summer the Arctic hare is gray, but in fall the animal molts into a beautiful pure-white winter coat. It loses its white coat in spring by molting again and often rolls in the snow to dislodge loose hair. The molted hair is left lying around in tufts on the tundra or snagged on the sparse vegetation, and it makes ideal nesting material for newly arrived migrant birds.

Eastern Cottontail

Sylvilagus floridanus

Eastern cottontails are widespread, common, and highly adaptable members of the rabbit family. They occur over a huge range, including most of the eastern and central United States.

Common name Eastern cottontail

Scientific name Sylvilagus floridanus

Family Leporidae

Order Lagomorpha

Size Length head/body: 11–13 in (29–34 cm); tail length: 0.5–1 in (1–3 cm)

Weight 2–4 lb (0.9–1.8 kg)

Key features Large rabbit with dense, soft, woolly gray-brown fur and white powder-puff tail; head small and rounded with long, oval ears

Habits Solitary and often bad tempered; active mostly at night

Breeding Three to 4 litters of 2–12 (usually 3–6) young born spring to early fall after gestation period of 26–32 days. Weaned at 16–22 days; sexually mature at 2–3 months. May live up to 10 years in captivity, up to 5 in the wild, but rarely more than 3

Voice Generally silent; may squeal in distress or alarm; females sometimes grunt in warning

Diet Grasses and herbaceous plants in summer; twigs, buds, and bark in winter

Habitat Varied; forests, swamps, meadows, and scrubland

Distribution Eastern North America from southern Manitoba, Ontario, and Quebec in Canada through eastern and midwestern U.S. and Central America to Colombia and Venezuela

Status Population: abundant

THE EASTERN COTTONTAIL THRIVES despite centuries of hunting for sport as well as for its meat and fur. The fur is warm, soft, and slightly shaggier than in other rabbits. It stays a grayish-brown color all year round, peppered with black guard hairs. There are one or two molts in a year, but cottontails never turn white in winter. Cottontails have been introduced for sport and meat to areas outside their natural range, including Washington State and Oregon. In other places they are regarded as pests because of the damage they can do to crops and ornamental plants. The key to the species' success is its adaptability. Eastern cottontails do well in all kinds of habitat, including forests, farmland, scrub, and even in cities, where they graze the vegetation in parks and gardens.

Antisocial Rabbits

In stark contrast to the European rabbit, cottontails are antisocial and highly intolerant of one another, so they are normally only seen one at a time. However, they are not fully territorial, and the ranges of adults usually partly overlap with those of other animals of either sex. Male home ranges are larger than those of females, and bucks generally reserve their aggression for fighting each other to gain dominance. Only females ever seem to actively defend any part of their range, and even then they are usually only concerned with the area immediately surrounding their nest.

Scent-marking helps define social boundaries, and secretions from glands in the cheeks are daubed on plant stems and other structures. Scent also plays an important part in

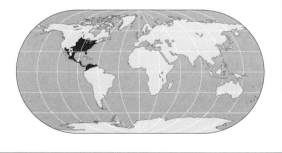

⊕ Although the eastern cottontail molts twice each year, it does not turn white in the winter, like the Arctic or snowshoe hare.

courtship. Vocal communication is limited to high-pitched alarm calls and coughing grunts. Cottontails are also known to produce high-pitched sounds by grinding their teeth together. This ultrasound is well outside the range of human hearing, but it is thought to be used by mother cottontails to reassure their young.

Rough Courtship

Courtship may last several days, during which the dominant male cottontail pursues a female. At first she keeps him at a distance, eventually signaling her acceptance with nuzzling, "kissing," and licking. Mating can be rather rough, with the male biting the female's neck as he mounts, sometimes inflicting serious wounds before being kicked away.

As the time for the litter to be born approaches, the female cottontail prepares several nests. Often they are in self-excavated bowl-like depressions in the ground—more like the forms of hares than the burrows of European rabbits. The nest the mother cottontail decides on depends on various environmental and social factors. She may even change her mind if she feels her family is under threat, moving them one by one to a new site. Baby cottontails are born small and utterly helpless. Nevertheless, the mother leaves them alone in the nest for most of the time, only returning to suckle them once a day until they are weaned at about three weeks. After that the young kittens may remain together for a further month before squabbles break out, and they develop the antisocial tendencies of adult animals. Cottontails born in the spring will have the opportunity to rear litters of their own before the summer is over.

Litter sizes vary throughout the season and from place to place. As a general rule, females living in the north produce larger litters than those in the tropics, but do so less often. The breeding season in the north is short, and there is usually only time for two or three litters to be raised in a year. In Central and South America four or five litters are the norm, but they often contain only two or three young.

Under Threat

Cottontails are prolific breeders—and need to be in order to compensate for heavy losses due to predation. The species serves as food for a wide variety of other mammals, as well as for some birds and reptiles. Most populations suffer an average of about 80 percent mortality a year despite many adaptations to avoid being caught. Cottontails are alert and wary, pausing often in their nightly foraging activities to survey their surroundings. Even when feeding, their eyes and ears swivel this way and that. The eyes are large, slightly bulging, and situated high on the sides of the head, giving almost 360-degree vision. The arrangement is excellent at detecting movement in the periphery, but not so effective for judging distance.

The cottontail's first response to a perceived threat is to freeze in the hope that its brown fur will prove effective camouflage. Individuals can stay perfectly still for a quarter of an hour or more, during which time the heart rate

↱ A cottontail in its nest, which is rather similar to the forms of hares. Baby cottontails will live in a separate nest from their mother; she will only visit in order to feed the kittens.

↓ An eastern cottontail suns itself near a burrow taken from a prairie dog. The eastern cottontail does not dig burrows itself, preferring to use those of other species.

Meet the Relatives

The eastern cottontail is one of 14 New World rabbits in the genus *Sylvilagus*. A closely related fifteenth species, the pygmy rabbit (*Brachylagus idahoensis*), is also sometimes included in the group. The various cottontails look quite similar but differ strikingly in habitat and ecology. The New England cottontail (*S. transitionalis*) is a forest dweller, and the desert cottontail (*S. audubonii)* and swamp cottontail (*S. aquaticus*) are adapted for life in arid and marshy areas respectively. Their specializations are reflected in much more limited distributions. Several species are threatened with extinction, including the Mexican Guerrero/Omilteneor cottontail (*S. insonus*) and the Dice's cottontail (*S. dicei*) of Panama and Costa Rica.

slows. It is the opposite to the escape response, in which a surge of adrenalin raises heart rate and primes muscles for action. Fleeing is a last resort—a cottontail will move suddenly, making two or three huge leaps in quick succession to confuse the predator. It then heads for cover; but instead of taking the most direct route, it bounds in a zigzag path, only adopting a flat run if it has a reasonable head start.

Disease Carriers

Like other *Sylvilagus* rabbits, eastern cottontails are naturally resistant to myxomatosis, the disease that devastated populations of the European rabbit when the *Myxoma* virus was deliberately introduced to Australia and Britain in the 1950s. However, cottontails suffer from tularemia, a bacterial disease that can be passed to humans through contact with infected meat or carcasses.

Cottontails are often regarded as pests of farms, orchards, and gardens. Most of the damage to trees is done during the winter when green fodder is in short supply, and grass is buried under snow. The cottontail's winter diet contains less water than the succulent green vegetation eaten in summer, and the animals will sometimes drink from puddles or even eat snow to make up the difference.

Common name American pika
(common pika, Rocky
Mountain pika, calling hare, mouse hare,
coney)

Scientific name *Ochotona princeps*

Family	Ochotonidae
Order	Lagomorpha
Size	Length head/body: 6.5–8.5 in (16–22 cm)

Weight 4–6.5 oz (110–180 g)

Key features Compact, rounded body
with short neck and short legs; no visible tail;
blunt head has short, round ears; fur is
grayish-brown to buff with buffy underparts

Habits	Solitary; aggressively territorial; terrestrial; diurnal; does not hibernate, but stores food for winter
Breeding	Two litters of 1–6 (usually 3) young born spring and summer after gestation period of 30 days. Weaned at 30 days; sexually mature at 3 months. May live up to 7 years in captivity, rarely more than 5 in the wild
Voice	Sharp alarm whistle and shrill courtship "song"; juveniles squeak
Diet	Green plant material supplemented with stockpiled hay during winter
Habitat	Boulder-strewn mountain slopes (talus) with patches of alpine grassland
Distribution	Mountainous parts of western Canada and U.S. from British Columbia to central California and Colorado
Status	Population: abundant

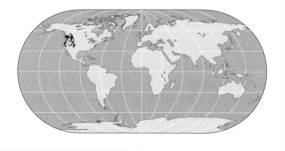

American Pika

Ochotona princeps

The common American pika is a charming and hardy little animal. Unlike most of its Eurasian relatives, it is also fiercely territorial: It will fight viciously to defend its home range, especially its painstakingly gathered winter larder.

THE AMERICAN PIKA IS SOMETIMES known as the common pika or Rocky Mountain pika in order to distinguish it from a second North American species, the collared pika (*Ochotona collaris*). The latter lives farther north than the common pika and endures some of the harshest winter conditions of any small mammal in Alaska and northwestern Canada.

Cute, but Not Cuddly

All pikas look rather similar, but the two American species can be distinguished by color. The American pika has buffy underparts, while those of the collared pika are almost white. They are undeniably cute-looking animals, with a short baby face, bright, beady black eyes, and a twitchy nose with long, straight whiskers. However, appearances can be deceptive and American pikas are usually anything but cute in their behavior toward each other. Unlike most Eurasian pikas, which live in family groups or colonies, American pikas are solitary and aggressively territorial.

Good pika habitat contains two vital components: an area of talus (boulders and broken rock) in which the pikas can shelter and store hay, and an adjacent foraging area, usually alpine meadow. Some populations survive on the edges of upland forests and climb into the lower branches of trees and shrubs to collect green leaves. The average territory is about 100 feet (30 m) in diameter, but range size and population density vary depending on the quality of habitat.

Pikas do not hibernate, and so they must continue to feed all year round. In winter, when most of their habitat lies under a thick blanket

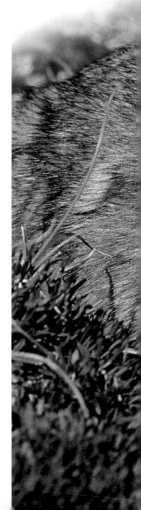

⊕ *Despite its cute looks, the American pika is an aggressive animal. Resident males will vigorously attack males straying onto their territory. Neighboring males use scent and calls to avoid each other.*

of snow, it can be difficult to find enough fresh food. The pikas use tunnels under the snowdrifts to visit foraging areas close to their homes. However, collecting food in this way is hard work, and there is often not enough to sustain them. To compensate for the shortfall, the pikas collect extra food in the summer and stockpile it for the lean times ahead.

Noisy Creatures

For such small animals pikas can be extremely noisy. The warning call is a sharp whistle, normally used when a bird of prey is seen. The sound sends every pika in earshot diving for cover. A similar call is used in territorial defense. While most pikas live in remote areas and are not used to people, those living in national

⊕ *A pika calls on top of a rock. Pikas have two calls—the short call and the long call, or "song."*

parks and other popular recreation areas can become bold and easy to watch. They often appear to be torn between the urge to run away and the tempting possibility of extra food offered from a picnic lunch. As a result, they remain where they are, but continue to sound the alarm, only scuttling away or whisking into a rocky crevice to safety at the very last minute.

A different kind of call, more of a song, is produced by males during the breeding season. It is the only time a pika will tolerate the presence of another within its territory. Females usually mate with a male they already know, one whose usual territory borders her own. The nonbreeding territories of males and females are roughly the same size; but once the male has been accepted as a mate, he also helps defend the female's home patch. His presence is tolerated throughout the rearing of two litters; but as the late summer hay-making season approaches, the female becomes increasingly antagonistic, and the male retreats back to his own territory.

Building a Haystack

American pikas are at their most aggressive during the haying season, when each animal builds a store of food to last it through the winter. The main haystack is usually built in the middle of the territory and is jealously guarded. Both males and females will attack intruders without hesitation, often inflicting serious injuries with their teeth. Even the female's own young will be ruthlessly expelled as soon as they are old enough to fend for themselves, at about one month of age. On leaving their mother's home range, juvenile pikas immediately set about looking for a vacant territory and claiming it for themselves. Most are able to settle within 100 yards (90 m) of the place where they were born. By three months the juveniles have reached full adult size and are capable of defending a territory, but they do not breed until their second year.

Pikas and People

American pikas do not reproduce anything like as fast as some Eurasian species, and consequently their populations do not suffer from the same wild fluctuations. Even in areas where pikas are numerous, they rarely cause nuisance to people because human habitations are few. Unfortunately, in areas where pikas and people live alongside each other, the pikas almost always come off worst, and several local subspecies have been reduced to critically low numbers. Recent estimates suggest that in several locations in Idaho, Utah, Nevada, New Mexico, and Wyoming the pika populations number fewer than 1,000 individuals. Because the animals live on isolated patches of land, there is little chance of recolonization if they should die out. However, most large American pika populations live well away from human habitation in remote and rugged areas of wilderness, much of which is protected inside national parks and nature reserves.

⊖ An American pika sits in its den in Mount Rainier National Park. The den is a place to shelter and store food for the winter. Many pikas have a tendency to store far more leaves and grass than they will consume over the cold months.

Making Hay while the Sun Shines

In late summer pikas begin collecting lots of fresh green grass and other herbaceous vegetation to make hay. The fresh material is transported from the foraging area one mouthful at a time and carefully laid out to dry in the sun. The resulting hay keeps much better than undried material and will still be perfectly edible many months later. As the summer progresses, pikas spend more and more time adding to their winter larder, sometimes collecting a dozen or more loads in an hour. Each batch of fresh hay is added to a personal stack, which by the onset of winter can weigh several pounds.

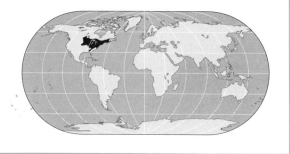

Common name American short-tailed shrew (northern short-tailed shrew)

Scientific name *Blarina brevicauda*

Family Soricidae

Order Insectivora

Size Length head/body: 3.5–4 in (9–10 cm); tail length: about 1 in (2.5–3 cm)

Weight 0.6–1 oz (17–28 g)

Key features Typical red-toothed shrew with a distinctly short tail; general color slate gray to almost black; underparts slightly paler

Habits Burrows in soft soil and leaf litter; rarely found on the surface; active day and night, punctuated by short rest periods

Breeding Litters of usually 4–7 young born February–September after gestation period of 21–22 days. Weaned at 25 days; sexually mature at 47 days. May live up to 33 months in captivity, up to 30 in the wild

Voice Quiet twitter; aggressive squeaks and shrieks if attacked

Diet Soil invertebrates, including small worms, insect larvae, and beetles; some plant material and occasional vertebrate prey

Habitat Dense ground cover in woodland and thick grassy areas

Distribution Much of northern-central and northeastern U.S.; southern regions of adjacent Canadian provinces

Status Population: unknown, likely to be millions. Abundant

American Short-Tailed Shrew

Blarina brevicauda

The American short-tailed shrew is one of the common small mammals of the northeastern United States, easily recognized by its long, pointed nose, red teeth, and noticeably short tail.

ALL THREE SPECIES OF SHORT-TAILED shrew forming the genus *Blarina* have a tail that is only about one-third the length of the head and body. This contrasts with the various species of long-tailed shrews, whose tail is about half the length of the body or longer. The American short-tailed shrew is also larger than any other shrew found in the northeastern United States and adjacent parts of Canada. The other two species of short-tailed shrews—the southern and Elliot's short-tailed shrews—occupy geographically separate areas of the United States and also have different numbers of chromosomes.

Burrowing Shrew

The short-tailed shrew is one of the common species of small mammals found in woodland and old grassland. It prefers habitat where there is dense ground cover, and it is rarely seen out in the open. The short-tailed shrew actually spends most of its time underground, often using the tunnels excavated by moles and other small mammals. But the species also digs its own burrows, and it is the most fossorial of the American shrews. Its tunnels may honeycomb large areas of soft soil and collapsed tangles of grass and weeds. The tunnels are generally within 5 inches (12 cm) of the surface, and like those dug by moles, they act as traps into which soil invertebrates fall. They will then be found and eaten by the shrew as it patrols its tunnel system. Digging shrews use their front feet for excavating, and the dirt is kicked away with the hind feet. Nests are built underground, usually of shredded grass and leaves, but they sometimes incorporate fur. The short-tailed

shrews do not hibernate and are active all winter, often using burrows under snow. At that time their food requirements increase substantially to over half their body weight daily.

The shrews forage for soil invertebrates, particularly small worms, millipedes, spiders, mollusks, and insect larvae. However, they will kill larger things, assisted by their ability to deliver a poisonous bite. The poison is useful in paralyzing large invertebrates, allowing them to be saved and eaten later. Surplus food is cached in the shrew's burrow, and the animal will return repeatedly to its store to consume what is there and to replenish supplies. The behavior is most common in winter, but also occurs in summer.

Plant material has sometimes been found in the stomachs of short-tailed shrews, but it may have been eaten accidentally along with animal prey, rather than deliberately consumed. For example, shrews attracted to rotting fruit or mushrooms for the maggots they contain would find it difficult to extract the animal food without swallowing some of the vegetable matter. They probably do not need to drink, since their food, particularly juicy earthworms, contains a lot of water. The shrews also lick

shrew can also climb well and may occasionally be found up in trees. Here it may perform a useful service by eating forest pests such as larch sawflies and other destructive insects.

Ravenous Appetites

Short-tailed shrews starve if they do not feed frequently. As a result, they need to be active throughout the day and night to fulfill their high energy requirements. They are active in brief bursts averaging four to five minutes, followed by rest periods lasting about 20 to 25 minutes. They seem to be busier in the early morning and also more active on cloudy days than in sunny or rainy weather. Short-tailed

⊕ *The short-tailed shrew can deliver a poisonous bite. The venom is strong enough to kill small animals, which ordinarily would be too large for the shrew to prey on.*

dew off the vegetation. Moreover, living in the humid air of tunnels in damp soil, and among moist grass, the shrews are likely to lose less water by evaporation from their body and lungs than if they lived out in the open.

Shrews have large scent glands on their flanks and bellies. They produce a strong odor that marks out their tunnels. The size of the glands and production of scent vary with the sex and reproductive status of the individual. Breeding males have the largest glands and produce the greatest amount of scent. The short-tailed shrew is said to use vocal clicks to detect objects by echolocation. That ability might help compensate for the fact that its eyes are smaller than pinheads and barely able to distinguish light from dark. In any case, eyes are of little use in underground burrows. Echolocation by shrews is not sufficiently sophisticated to detect small objects the way that bats can, but it should help locate tunnel openings and tell the difference between those that are blocked and routes that are accessible.

Young Wanderers

While adults seem to have a fixed home range averaging about 5 to 6 acres (2 to 2.5 ha), the younger animals may be nomadic until they find a place for themselves. Short-tailed shrews are solitary animals—particularly the males and

Toxic Saliva

The saliva of the short-tailed shrew is poisonous and contains chemicals that paralyze the nervous system (neurotoxins) and damage blood cells (hemotoxins). It is not dangerous to large animals such as humans, although a bite can remain painful for several days. The toxic saliva is similar to the venom produced by some snakes. However, unlike snakes, which have hollow fangs, the shrew cannot inject the poison. Instead, it is chewed into the wound at every bite. The toxin is sufficiently effective that it will kill mice and other large prey, enabling the shrew to prey on larger animals than would be expected. The saliva will also paralyze invertebrates such as mollusks, preventing their escape and allowing them to be eaten later.

older individuals who tend not to be sociable at all. Population densities vary from year to year and range from less than one to 50 per acre (120 per ha). Dense populations may crash late in the year and take several years to recover. Winter mortality may reach 90 percent, especially in bitterly cold weather.

Rough and Tumble

The breeding season begins in early February (later farther north) and lasts until September, with the females tending to come into breeding condition earlier than males. Mating is a rough and tumble affair, with the shrews joined together for several minutes. During that time

family then disperses. The juveniles are capable of breeding when they are under two months old. Adult females can give birth to as many as three litters in a season, so the shrew population often builds up rapidly and reaches high numbers by the end of the breeding season in September.

Natural and Man-Made Perils

The short-tailed shrew, like many other small mammals, faces a wide range of predators. Owls are a particular threat, since they have a poorly developed sense of taste and are not put off by the shrew's strong-smelling skin glands. Consequently, the short-tailed shrew is a frequent victim. Shrews are also highly inquisitive animals and will push into small spaces to seek food. As a result, they often fall victim to discarded bottles and drink cans, unable to escape having squeezed inside.

A subspecies of the short-tailed shrew, found only at one place in southwestern Florida, may have been wiped out by development of its habitat and predation by domestic cats. The many threats faced by shrews ensure that only 11 percent live more than one year, and very few reach two years old. That is partly because their gritty food, such as earthworms, wears down their teeth, which do not grow continually like the incisor teeth of rodents. As a result, old shrews are less able to chew their food properly, even though they have 32 teeth. The animals are also likely to get worn out by having to be active day and night throughout their lives. Average survival is only about four or five months.

In October and November the shrews shed their summer coat and replace it with longer winter fur. The winter molt begins on the tail and moves forward toward the head. The spring molt begins in about February, but can take place any time up until the early summer. The dark-gray winter coat is then replaced by shorter, slightly paler fur. This time the molt starts at the head and proceeds toward the tail. Why the molt should go in different directions at different seasons is a mystery.

the female continues to move around, dragging the male behind her. Females seem to need several matings in a day in order to induce their ovaries to release eggs that are then fertilized and develop directly into embryos. Gestation lasts for three weeks, after which the young are born blind, pink, and helpless. The normal family size is between four and seven. The newborn young are hairless except for their whiskers. They take 25 days to be weaned and are fully independent within four weeks. The

⊕ A short-tailed shrew feeds on the carrion of a meadow vole. The shrews need to hunt for prey at all times of day and night, only stopping to rest for brief periods.

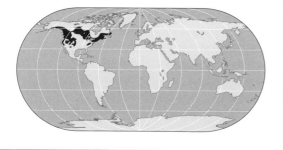

American Water Shrew

Sorex palustris

The American water shrew can actually walk on water, helped by stiff hairs that trap bubbles of air under each foot, allowing the shrew to run quickly over the surface of a pool.

Common name
American water shrew

Scientific name *Sorex palustris*

Family Soricidae

Order Insectivora

Size Length head/body: 3–4 in (7.5–10.5 cm); tail length: 2.5–3 in (6–7.5 cm)

Weight 0.3–0.5 oz (8.5–14 g)

Key features Relatively large shrew; blackish-gray coat sometimes becoming browner in summer, pale to dark-gray underside; 2-colored tail; distinctive fringe of stiff white hairs on sides of feet; tiny eyes and ears; red tips to teeth

Habits Solitary; active day and night but mostly just after sunset and before dawn; hunts in water

Breeding Litters of 3–10 young born late February–June after gestation period of about 21 days. Weaned at 28 days; sexually mature at 2 months in early-born young, 10 months in late-born young. May live about 18 months in captivity, similar in the wild

Voice High-pitched squeaks during territorial disputes and habitat explorations

Diet Aquatic invertebrates such as caddis fly larvae and insect nymphs; occasionally small fish; on land takes flies, earthworms, and snails

Habitat Waterside habitats from bogs to fast-flowing mountain streams, especially in northern forests; prefers humid conditions

Distribution Canada; southeastern Alaska; mountain regions of U.S. into Utah and New Mexico, Sierra Nevada to California

Status Population: unknown, but likely to be millions. Widespread and abundant

WHEN FORAGING FOR PREY underwater, the American water shrew's body fur traps a layer of air. This helps keep it warm in the water, but also gives the shrew extra buoyancy. The shrew requires more effort to prevent itself bobbing to the surface, and it has to swim vigorously to keep itself submerged. The shrew has special stiff hairs on its feet that increase the resistance to water, making them more efficient as paddles. The water shrew will swim, dive, float, and run along the bottom of a stream in its search for food. The animals are also able to hunt and catch small fish, demonstrating their remarkable agility in the water.

Once they have captured their prey, it is held in the forepaws and torn to pieces using the shrew's sharp teeth and upward-tugging motion of the head. Vegetation found in the stomachs of some water shrews may have been eaten accidentally with the usual insect prey.

Regulating Their Temperature

The American water shrew has the ability to control the way blood flows around its body so that it can even dive during the winter. It keeps from freezing to death by diverting blood away from the surface of the skin, where it would quickly get cold. Each dive can last for up to 48 seconds—a very long time for such a tiny creature to hold its breath.

The shrew manages this feat by using its oxygen supply to power the muscles and slowing down other body processes such as the heartbeat and digestion. Immediately after swimming, the shrew dries off its coat, using its hind feet to brush away water from its fur.

It has been suggested that water shrews may be able to echolocate as bats do, since they make constant high-pitched squeaks as they explore their territory. However, there is little evidence to support this, and they certainly cannot use sound in such a precise way as bats. They probably mainly use their sense of smell to locate prey on land and their whiskers to detect movements of prey when underwater.

Never Satisfied

American water shrews can go without food for up to three hours, but they usually feed far more frequently—every 10 minutes on average during active periods. Every day they must eat at least five to 10 percent of their body weight. To find enough food, shrews hunt over territories of up to 1 acre (0.4 ha).

Water shrews are very aggressive toward any other shrews of either sex. When two shrews meet, they will squeak a warning and stand on their hind legs to show their pale-colored stomachs. If neither shrew backs down, there will be a fight in which the shrews wrap up into a tight ball, biting each other. Tail and head injuries are common from such encounters.

The only time a shrew will tolerate another is to mate; although once mating has taken place, the female will chase off the male and raise her family alone. It is thought that mates may be attracted by the strong, sometimes nauseatingly powerful odor emitted by the shrew. The water shrew's nest is about 6 to 8 inches (15 to 20 cm) in diameter, built under a boulder, inside a hollow log, or in a tunnel dug by the shrew. Shrews dig as a dog would, scraping at the soil with their front feet and pushing it back with their hind feet. The nest of shredded grass, leaves, and dried vegetation is then pushed into place with the muzzle.

The scientific name of the American water shrew comes from the words *soric,* meaning "shrew-mouse," and *paluster,* meaning "marshy"—a good description of where it normally lives at the water's edge. It particularly favors mountain streams up to 4,000 feet (1,200 m) above sea level. Here the water is cool, has plenty of dissolved oxygen, and therefore supports abundant aquatic invertebrates on which the shrew can feed. The American water shrew is a relatively large, typical long-tailed shrew. The tail is longer than the body, which is unusual in the shrew family. It probably helps the water shrew in steering when it swims. It has many predators, which on land include owls, hawks, opossums, foxes, bobcats, weasels, and skunks. In the water it faces another range of hungry attackers, including large fish such as trout, and garter snakes. The water shrew must often use its swimming ability to escape from such predators.

⊖ *An American water shrew on the shore of a Colorado creek. The species' scientific name literally means "marshy shrew-mouse."*

Star-Nosed Mole

Condylura cristata

The star-nosed mole looks quite unlike any other mammal. It boasts a bunch of 22 fleshy projections encircling its nostrils, appearing like a pink flower at the end of the animal's nose.

Common name Star-nosed mole

Scientific name *Condylura cristata*

Family Talpidae

Order Insectivora

Size Length head/body: 4–5 in (10–13 cm); tail length: 2–3 in (6–8.5 cm)

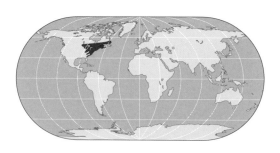

Weight 1.4–2.9 oz (40–84 g)

Key features Black animal with large hands; characteristic cluster of fingerlike fleshy projections around the end of nose

Habits Digs tunnel systems with small piles of dirt ejected to the surface at intervals; active throughout the year and all times of the day and night

Breeding Single litter of 3–7 young born April–June after gestation period of probably about 3 weeks. Weaned at around 3 weeks; sexually mature at 10 months. May live about 5 years

Voice Normally silent

Diet Large soil invertebrates, mainly earthworms, but also fish and aquatic insect larvae

Habitat Damp or muddy soil in fields, meadows, woods, and marshy areas

Distribution Northeastern U.S. south to Georgia; southeastern Canada from Labrador to Manitoba

Status Population: locally abundant in the north, but scattered and less common in the south

STAR-NOSED MOLES LIVE IN DAMP and very wet soil, in contrast to other mole species that often prefer drier places. They create tunnels about 2 inches (5 cm) in diameter by compressing the earth to the sides or pushing it up as a ridge on the surface of the ground. They also dig burrows by scraping away earth with their very large, square hands and pushing it to the surface at intervals to form molehills. They are about 12 to 24 inches (30 to 60 cm) wide and can be up to 6 inches (15 cm) high. In wet ground the water table (the level at which the soil is completely saturated, and burrows will become flooded) is quite near the surface and limits how deep the mole can dig. Mostly the network of tunnels is just below the surface, but in drier ground burrows may be as much as 24 inches (60 cm) deep. The burrow system has a number of chambers, at least one of which will be furnished with a nest made of dry vegetation about 6 inches (15 cm) in diameter. It is normally constructed above the level at which the soil is saturated, sometimes even above the surface of the ground, but safely tucked away underneath a log or tree stump and linked by tunnels to the main burrow system.

The star-nosed mole patrols its network of tunnels in search of food in the form of soil invertebrates that fall into the underground passageways. They include earthworms, insect larvae, and small mollusks. Although most of its time is spent underground, the star-nosed mole also comes to the surface at night and often forages extensively there, particularly in search of worms that lie around in damp grass on moist summer evenings.

Indigestible Diet

Many of the invertebrates that the moles eat have a high proportion of their body made up of indigestible materials such as shells and exoskeletons. Worms have a very high water content and also contain large amounts of dirt. Consequently, the star-nosed mole has to eat large quantities of that kind of food in order to get enough nourishment from its diet, which requires the mole to be active both day and night and also throughout the year.

During the winter it may construct a series of tunnels under the snow that enable it to forage on the ground surface without being exposed to predators. Part of the tunnel system is often

⊕ The star-nosed mole sometimes comes to the surface at night to forage for worms lying in the damp grass.

flooded, and there is usually at least one tunnel entrance that opens underwater. Living in such wet places, the mole needs its waterproof fur in order to keep warm. The star-nosed mole is at home in the water, and it is also a very competent swimmer, using all four feet to propel itself. In winter it will dive and swim under the ice. Diving enables the mole to obtain food from the bottom of ponds and streams. It can remain submerged for long enough to search for aquatic insects, small fish, and crustaceans. They represent an additional source of nourishment to food that the mole finds in its own tunnels. The peculiar fingerlike fleshy appendages at the end of the snout are used to feel for and recognize prey. These sensitive projections can be used to help manipulate food. They are constantly moving, except for the two middle ones on top of the snout, which point forward like fingers. These additional sensory structures on the snout of

Peculiar Projections

The star-nosed mole's scientific name, *Condylura cristata*, draws attention to the peculiar structures on the nose—*crista* meaning "crest" or tuft. The fleshy fingerlike projections are extremely sensitive and help the animal feel for its prey, both in the darkness of the burrow system and also underwater. The sensitivity is partly due to large numbers of so-called Eimer's organs, like microscopic goose pimples on the surface of the thin skin. The projections around the end of the snout are not just sensitive to touch, but may also be able to detect minute electrical disturbances in the water created by muscle and nerve activity in the mole's prey. The projections are also capable of being moved and can be used to help manipulate the mole's food. The name *Condylura*, meaning "lumpy tail," is a reference to an old illustration of the animal that showed a knobby tail. However, live animals do not have a lumpy tail, and it is possible that the drawing was made from a badly preserved animal in which the tail had dried into a lumpy condition.

the star-nosed mole must be very useful, since its eyes and ears are tiny and buried among the dense fur. Eyes especially are of little use underground—an enhanced sense of touch is much more help.

Shared Burrows

Moles are generally solitary animals, each one occupying a home range covering about 0.9 acres (0.4 ha). However, unlike most other species of mole, the male and female star-nosed mole may live together, sharing the same burrow system at least during the winter.

The summer coat is molted in September or October, and a dense winter coat replaces it. At this time of year the tails of both sexes become enlarged with stored fat. It probably serves as an energy reserve to help tide the animals over during periods of food shortage.

Litters are mostly born in May or June, but can be as early as March or as late as August. The young are born into a nest made of dry grass and leaves collected at the surface. Breeding nests are generally larger and more elaborate than those used at other times. At birth the young are naked except for a few short whiskers on the snout, but they soon grow a covering of fine hair. They produce their first young in the spring following their birth. Generally, the population density is about one per acre (2 per ha), but in good swampy habitats there may be five times that number.

The species spends more time on the surface than is usual among moles, exposing it to greater risk of predation. A variety of carnivores eat star-nosed moles, ranging from house cats to snakes, and even large-mouth bass. Generally speaking, star-nosed moles do not encroach on human activities. That is partly because their home, low-lying wet ground, is not an economically important habitat. However, where the moles invade golf courses or lawns, their molehills are unwelcome.

⊕ *The star-nosed mole's unique nose is divided into a number of fleshy tentacles that are used for locating prey both in the soil and underwater.*

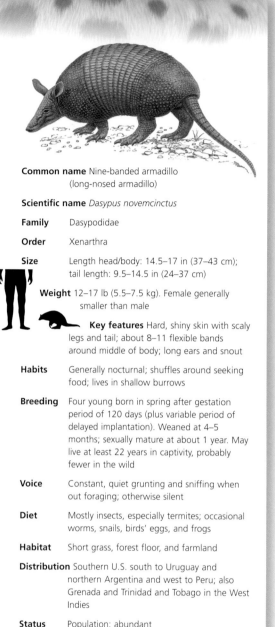

Common name Nine-banded armadillo
(long-nosed armadillo)

Scientific name *Dasypus novemcinctus*

Family Dasypodidae

Order Xenarthra

Size Length head/body: 14.5–17 in (37–43 cm);
tail length: 9.5–14.5 in (24–37 cm)

Weight 12–17 lb (5.5–7.5 kg). Female generally
smaller than male

Key features Hard, shiny skin with scaly
legs and tail; about 8–11 flexible bands
around middle of body; long ears and snout

Habits Generally nocturnal; shuffles around seeking
food; lives in shallow burrows

Breeding Four young born in spring after gestation
period of 120 days (plus variable period of
delayed implantation). Weaned at 4–5
months; sexually mature at about 1 year. May
live at least 22 years in captivity, probably
fewer in the wild

Voice Constant, quiet grunting and sniffing when
out foraging; otherwise silent

Diet Mostly insects, especially termites; occasional
worms, snails, birds' eggs, and frogs

Habitat Short grass, forest floor, and farmland

Distribution Southern U.S. south to Uruguay and
northern Argentina and west to Peru; also
Grenada and Trinidad and Tobago in the West
Indies

Status Population: abundant

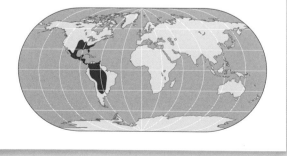

Nine-Banded Armadillo

Dasypus novemcinctus

*The armored armadillo is an unmistakable creature,
and the nine-banded species has been a successful
colonizer of the United States.*

INSTEAD OF HAVING A SOFT, FURRY skin like most
other mammals, the armadillo is encased in a
bony carapace that is covered with shiny plates.
Around the animal's middle is a series of flexible
bands that allow the creature to roll up when
threatened. Usually there are about eight such
bands, but armadillos living in Central America
have nine or more. The tail and legs are scaly,
and there are sparse yellowish hairs here and on
the animal's belly. The armadillo has large,
sensitive ears and a long snout. The powerful
front feet each have four toes, but the hind feet
have five. Despite its heavily armored body, the
armadillo is a strong swimmer and can remain
underwater for long periods. It can also run
surprisingly fast, and the smooth and shiny
body is difficult for predators to grasp.

Traveling North

The main home of armadillos is in South
America. The nine-banded species is widespread
there, being found as far south as Uruguay and
northern Argentina. In the last 200 years it has
also staged a remarkably successful invasion of
North America. From the late 1800s onward the
nine-banded armadillo rapidly expanded its
range in northern Mexico and had reached as
far north as the Rio Grande region of southern
Texas by 1890. Since then it has spread steadily
deeper into the United States, appearing in
Tennessee by the 1970s. It is now found as far
north as Nebraska and southern Missouri.
Meanwhile, in the 1920s armadillos were
released or escaped from captivity in Florida in
several places. The animal has now become well
established there and has spread north and
west to colonize the southern states of the

United States, reaching Georgia and South Carolina by the 1950s. It is now sufficiently numerous in Florida to have become one of the more frequent victims of road traffic accidents.

It is not clear what has brought about the rapid extension of the armadillo's geographical range, but climatic changes might have something to do with it. More significantly perhaps, the persecution and removal of large predators such as cougars leaves the way open for the armadillo to go where it pleases, with the smaller carnivores posing little danger. Another helpful factor might be the steady expansion of ranching. Overgrazing by cattle leaves the grass short and the soil nicely warmed by the sun. Such conditions are ideal for many ground invertebrates, and the short grass enables armadillos to find them easily. Further spread to the north is now probably limited by the cold, especially in winter. In many parts of the western states the summers are also too hot and dry.

The armadillo is an unusual-looking creature. It has large, rounded ears, a long snout, and is covered in hard, shiny armor plating.

Well Protected

In the warmer parts of its range the nine-banded armadillo prefers to live in shady cover. Elsewhere, it thrives in various habitats from sea level to altitudes of more than 10,000 feet (3,000 m). The armadillo is mainly active in the late evening and at night, although it may sometimes come out during daylight on cloudy days. It shuffles around slowly, nose to the ground as it sniffs for prey. When rooting, the animal grunts almost constantly, ignoring the potential danger of drawing attention to itself. But it is well protected by its thick, bony skin. The animal can also roll up to hide its softer underbelly. Being relatively safe from predators means the armadillo can potter around confidently, even out in the open.

Periodically the armadillo will rear up on its hind legs, supported by the tail, and sniff the air. The animal has a keen sense of smell and pokes its nose into clumps of dry vegetation and leaf litter, seeking out food. It will often pause to dig something up, using the strong claws on its forefeet, and it may also rip apart rotting logs. Insects make up more than three-quarters of the diet, with termites a frequent item on the menu. Even the nests of fire ants may be attacked to get at the ant larvae within. The armadillo's thick, bony skin protects it from painful insect bites. The animal will also eat small worms, mollusks, and occasionally the eggs of ground-nesting birds.

Unimpressive Fighters

The armadillo's home range extends over about 3 to 4 acres (1.5 ha) in good habitat, but in poorer areas it may cover more than 30 acres (12 ha). The animals are tolerant of each other, and their home ranges often overlap. However, at high population densities the animals may become less accommodating, and the males will fight, scratching each other with their front feet. Armadillos are unlikely to cause serious harm and do not bite because their teeth are small and form only simple pegs. The jaws are weak too, since they are meant only for picking up small insects.

Armadillos excavate a burrow by digging with the forefeet and kicking loose dirt out of the tunnel with the hind feet. The burrow is

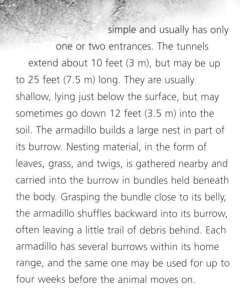

simple and usually has only one or two entrances. The tunnels extend about 10 feet (3 m), but may be up to 25 feet (7.5 m) long. They are usually shallow, lying just below the surface, but may sometimes go down 12 feet (3.5 m) into the soil. The armadillo builds a large nest in part of its burrow. Nesting material, in the form of leaves, grass, and twigs, is gathered nearby and carried into the burrow in bundles held beneath the body. Grasping the bundle close to its belly, the armadillo shuffles backward into its burrow, often leaving a little trail of debris behind. Each armadillo has several burrows within its home range, and the same one may be used for up to four weeks before the animal moves on.

Armadillos and Leprosy

In the 1960s it was discovered that nine-banded armadillos—unlike most other mammals—could be infected with the leprosy bacillus. For the first time the disfiguring disease could be studied in the laboratory. Leprosy was later found in wild armadillo populations in Texas and Louisiana, and (less often) in Florida. There is probably little risk of humans catching the disease from these wild animals, since leprosy is an uncommon disease, and people living outside the tropics tend to be less susceptible to leprosy.

young are born in March or April, but as early as February in Mexico. Litter sizes are small, normally four identical same-sex quadruplets derived from a single egg. Young armadillos are born fully formed with their eyes open. They weigh 1 to 2 ounces (28 to 56 g) and can walk within a few hours. They will accompany their mother on foraging expeditions within a few weeks and become fully independent at an age of four or five months.

Popular Food Source

Armadillos can become quite numerous in places and may reach densities of 130 per square mile (50 per sq. km) on the coastal prairies of Texas. In parts of South America they are a popular source of food, and catching them remains a threat to populations in many areas. The animals are also threatened by deforestation, agricultural expansion, and other forms of habitat loss. In the United States armadillos sometimes make themselves unpopular by digging in gardens and farmland. They are also accused of causing erosion and undermining buildings by their burrowing activities. However, on the whole armadillos are beneficial animals that destroy many harmful insects and are generally regarded with amusement and tolerance. Their bony skin is made into baskets, which are sold as souvenirs, and the animals are also used in various forms of medical research.

Occasionally, armadillos can be found nesting above the ground in large piles of dry vegetation.

Armadillos normally live alone, but meet to breed once a year in the summer months. (In captivity they may breed throughout the year.) They breed for the first time at about one year. Courtship is often a drawn-out affair, with the males eagerly following females and seeking an opportunity to mate. In North America the mating takes place in July and August, but implantation of the fertilized egg is delayed until November. Farther south mating may occur earlier in the summer, but implantation is then delayed for longer. Elsewhere, development of the embryos may start immediately after mating. Actual fetal development takes about 120 days. In Texas the

① A nine-banded armadillo excavates the earth by digging with its powerful clawed forefeet. Its body is well protected by bony armored skin.

⊙ A mother and infant nine-banded armadillo in a burrow. The armadillo builds a nest from leaves, twigs, and grass in part of its burrow.

Common name
Mexican free-tailed bat (guano bat, Brazilian free-tailed bat)

Scientific name *Tadarida brasiliensis*

Family   Molossidae

Order   Chiroptera

Size   Length head/body: 3.5–4 in (9–10 cm); tail length: 1–1.5 in (2.5–4 cm); forearm length: 1.5–2 in (4–5 cm); wingspan: 12–14 in (30–35 cm)

Weight 0.4–0.5 oz (11–14 g)

Key features Medium-sized bat with tail that projects well over tail membrane; ears broad; velvety reddish to black fur; wings long and narrow

Habits   Roosts in groups of up to 20 million individuals; flies high, hunting insects at night

Breeding   Single baby born in June after gestation period of about 90 days. Weaned at 6 weeks; sexually mature at about 1 year. Not kept in captivity, may live at least 18 years in the wild

Voice   Ultrasonic echolocation calls at 40–62 kHz; variety of social calls, including squeals, chirps, and buzzing audible to humans

Diet   Small insects such as mosquitoes, flies, beetles, and moths

Habitat   Wide variety of habitats from desert to pine and broad-leaved forest; roosts in caves, old mines, hollow trees, and buildings

Distribution U.S. to Chile, Argentina, and southern Brazil; Greater and Lesser Antilles

Status   Population: many millions; IUCN Lower Risk: near threatened. Believed to be fewer now than in recent past

Mexican Free-Tailed Bat

Tadarida brasiliensis

Mexican free-tailed bats can roost in huge crowds numbering over 20 million animals. Their prodigious appetite for migrating moths helps minimize pest damage to crops in southwestern America.

MEXICAN FREE-TAILS ARE medium-sized bats with short, velvety fur that can be reddish, dark brown, or almost black. Some individuals have white patches, which can be anywhere on the body. An almost mouselike tail gives the species its common name. Over half the length of the tail extends beyond the tail membrane that stretches between the legs. The ears are broad and black; and unlike most bats in the Molossidae family, they are not joined on top of the head. However, their upper lips have vertical wrinkles common to all bats in the family. The feet have distinct white bristles, used for grooming, on the sides of the toes.

Insect-Eaters

Mexican free-tailed bats are insectivorous; and like most chiropterans, they use echolocation to find and catch insects on the wing. Bats can eat their own weight in insects each night, with the appetite of nursing mothers being particularly demanding. A large bat colony might eat several hundred pounds (about 100 kg) of insects in a single night.

Although the first specimen to be described scientifically was from Brazil (hence the species name of *brasiliensis*), Mexican free-tailed bats are most common in Texas and Mexico. Their wide distribution stretches across the southern states of North America down through Central America, the Antilles, and most of western South America to central Chile and east to the coastal provinces of Brazil.

Mexican free-tails roost in caves and under bridges. They may sometimes use old mines, well shafts, and hollow trees. Because they

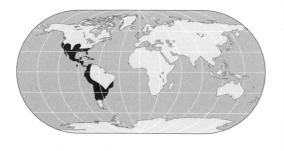

frequently occur in buildings they are often called "house bats." They usually roost near lakes and ponds where they can get drinking water, but mainly because of the insects that live around them. Although most of these roost sites will contain relatively small numbers of bats, some roosts are huge. There are caves in Texas where over 20 million bats congregate to have their young—that is more than the number of people in our largest cities. These bats form the largest known colonies of vertebrates anywhere in the world. At some famous caves, such as the Carlsbad Cavern in New Mexico, the dramatic sight of millions of

The tail of the Mexican free-tailed bat is thicker than in other bats. It does not lie in a broad tail membrane, but instead hangs free.

Mexican free-tails are insectivorous. They are known to feed on corn earworm moths, whose caterpillars are a major crop pest.

bats emerging like smoke from the cave entrance at sunset draws hundreds of spectators every night.

As well as the tourists, the bats also attract predators such as owls, hawks, and snakes, which wait at the cave mouth and snatch bats as they fly past. But with such huge numbers of bats, the predators can only catch a tiny fraction of the population.

Such huge numbers of bats create enormous mounds of waste. The levels of ammonia in the roost caves make the air stink. Concentrations can be so high that it is difficult to breathe and can even be lethal. The bats survive partly by producing large quantities of mucus in their airways. The stinking piles of guano (excrement) below the roosts are used as fertilizer, and sodium nitrate extracted from guano was once used to produce gunpowder.

The Mexican free-tailed bat clearly demonstrates one of the dilemmas of many bat species, especially those that feed on insects. The energy costs of flying and rearing fast-growing young are high. Furthermore, because of their relatively small size bats lose heat quickly, so they use a lot of energy in keeping warm. To compensate for their high energy needs, bats need to eat a lot; but in seasonal climates the numbers of flying insects fluctuate widely over the year. As the weather gets cooler, the insects disappear just when the bats' energy needs are increasing. Bats have two options: to sit out the cold period in torpor, using as few energy reserves as possible, or to fly to warmer, more productive climates.

In Mexican free-tails migration patterns vary depending on where the populations are based. Those in the northernmost part of their range in northern California do not migrate, perhaps because they would have to fly too far to reach warmth. During wintry periods the bats become semitorpid in cold roosts. However, unlike bats that go into true hibernation, their body temperature does not drop to match that of the surroundings. Instead, the bats cool to a few degrees below normal, saving substantial amounts of energy. Populations in Arizona do not migrate either, but spend winter resting in relatively warm roosts in chimneys, caves, and tunnels. Those in other southwestern states, such as Texas, migrate to Mexico. Migrating bats can travel distances of over 620 miles (1,000 km) in a couple of weeks.

A Bat Calendar

The populations that have been studied most closely are those migratory bats that fly south in winter to follow the warmth and insects. In spring they leave Mexico to fly to Texas. When they arrive, they mate. Males and females then separate into bachelor and nursery colonies. Little is known about mating behavior in the wild, but in captivity males claim territories and are aggressive toward intruding males. They mark their territory by rubbing their chests and scented throat glands on the roost surfaces. They "sing" to the females, who do the rounds, visiting different males' territories for mating.

Females give birth around June to a single naked pup. Most births in a roost fall within a 10-day period, so all the young are of a similar age. When the mothers fly off to feed, they leave their babies hanging huddled together in a "nursery patch" on one part of the roost. The mothers return once during the night to feed their young. When a mother arrives back, she is greeted by a barrage of youngsters all trying to steal a feed. Each female finds her own baby, first by using high-pitched "peep" contact calls, then by sniffing it to confirm its identity by smell. Sometimes a mother will feed a youngster that is not her own, but 85 percent of the time she manages to find and feed her own offspring among the thousands of others.

The appetite of the young bat is enormous. It can consume its own weight of milk every day, which means that the mother has to produce up to a quarter of her body weight in milk every 24 hours. In the warm environment of the nursery roost, kept even hotter by all the tiny bodies, babies grow quickly. Within a month most babies are furred and ready to fly. Males are sexually mature at 18 to 22 months, but females can produce their first youngster at one year old.

Farmers' Friends

A cloud of bats emerging from the mouth of a cave will climb high into the air until they are almost out of sight. Why they fly so high was once a mystery, but solving it has given a reason for farmers to respect these bats.

Radar studies of bats began at an airport, when the bats' movements were mapped to avoid collisions with planes. They have revealed amazing facts about the bats' flight patterns. The bats can fly at least as high as 10,000 feet (3,000 m), with groups showing foraging behavior at 600 to 3,200 feet (200 to 1,000 m). Microphones tied to weather balloons have recorded the feeding buzzes of free-tailed bats (the intense burst of echolocation used to pinpoint an insect just before it is captured) nearly half a mile high.

Although it seems improbable that bats can catch any insects a mile or more above the surface of the earth, they are actually tapping into the migration routes of one of the United States' most serious agricultural pests—the corn earworm moth. These insects gradually spread northward every spring from their winter

Swift Fliers

Free-tailed bats fly very fast and hunt insects in open spaces, particularly above water. They usually fly high, generally above 50 feet (15 m), unless they are swooping low over water to drink. Their efficient flight enables them to cover long distances on migrations. Their long, narrow wings and characteristic flight patterns resemble those of swifts. They are the high-speed bats, contrasting with the smaller, fluttery types found in woodlands and confined spaces.

strongholds in Mexico. Their caterpillars (also known as cotton bollworms) do much damage to crops. The adult moths fly hundreds of feet high into the air to catch the prevailing winds that carry them north in their seasonal migration. This way clouds of moths can travel over 100 miles (160 km) in one night. By intercepting the swarms, the bats eat moths that would otherwise cause damage as far north as Canada. A single colony of 20 million bats could eat nearly 15,000 tons (13,000 tonnes) of insects over one summer.

Insecticide Menace

Many bat colonies in the southern United States have shown a marked decline in numbers over recent years. The drop in numbers has been blamed partly on the use of the insecticide DDT. The bats eat contaminated insects, and the poison accumulates in the bats' bodies. Although DDT is now banned in the United States, if it is used on any part of the bat migration routes, the problem will not disappear. DDT also causes breeding failure, so the effect on the bat population can be serious.

Free-tails have one of the highest rabies infection rates among bats, so it is advisable to avoid handling them. However, the risk of being bitten by an infected bat is quite small.

⊖ *Mexican free-tailed bats emerge from Bracken Cave, Texas. In some places the spectacle of millions of bats appearing like a smokescreen draws crowds of tourists.*

271

Little Brown Bat

Myotis lucifugus

Little brown bats are common in urban and rural areas across America. In winter they may migrate considerable distances to find suitable caves for hibernation.

THE LITTLE BROWN BAT IS one of America's most common bats. It has been well studied because of its abundance and wide distribution. Its habit of roosting in houses makes it easy to examine.

Little brown bats are insectivorous. The bats hunt in flight, frequently foraging over water for insects such as mosquitoes, caddis flies, mayflies, and midges. They can catch an insect with the tip of the wing, transfer it onto the tail membrane, then curl their tail forward to scoop it into the mouth—all while flying in the dark at over 20 miles per hour (30 km/h).

Like other insectivorous bats that live in temperate regions of the world, the little brown bat's lifestyle is dictated by the seasons. It spends spring, summer, and fall rearing offspring and feeding, so that by winter it has enough fat reserves to survive winter hibernation. Females have to eat twice as much as males to cover the energy they expend on producing and nursing their young.

Second Homes

During the spring, summer, and fall bats use two types of roost—day roosts and night roosts. They rest in the day roosts, which can be in buildings, trees, under rocks, in piles of wood, and occasionally in caves. In the evening the bats gather in night roosts just before flying off to feed. The purpose of such nightly gatherings is not clear, but clustering together may help the bats raise their body temperature for the evening's flying. It also allows for social interactions and recognition of other members of the colony.

Common name
Little brown bat

Scientific name *Myotis lucifugus*

Family   Vespertilionidae

Order   Chiroptera

Size   Length head/body: 3–4 in (8–10 cm); forearm length: about 1.5 in (3.5–4 cm); wingspan: 9–11 in (22–27 cm)

Weight 0.25–0.45 oz (7–13 g)

Key features Small, fluttery bat with glossy fur of shades of brown; underside paler; small, black ears with short, rounded tragus; long hairs on toes

Habits   Hunts insects over water; females gather in summer nursery roosts, often in buildings; sexes hibernate together in caves over winter

Breeding   Single young born May–July after gestation period of 60 days. Weaned at 3 weeks; sexually mature in first year in southern parts of range, second year farther north. May live several years in captivity (not usually kept for long periods), 30 in the wild

Voice   Short pulse echolocation calls 38–78 kHz, peaking at about 40 kHz, too high-pitched for humans to hear

Diet   Insects caught on wing, especially mosquitoes, caddis flies, mayflies, and midges

Habitat   Urban and forested areas; summer roosts in buildings or under bridges, usually near water; hibernation roosts in caves or mines

Distribution Alaska to Labrador and Newfoundland (Canada) south to Distrito Federal (Mexico)

Status   Population: probably several million. Not seriously threatened

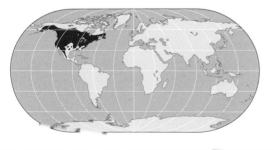

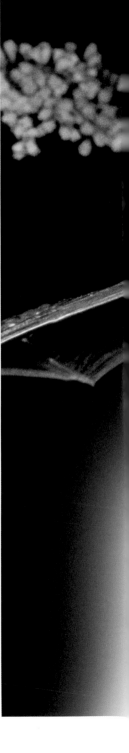

travel up to 300 miles (500 km) to find a suitable place. A single cave in Vermont, for example, draws in bats from all over New England. The bats hibernate in clusters. Their body temperature drops to almost that of their surroundings and slows their metabolic process to practically a standstill. At 35°F (1.6°C) the bat's oxygen consumption is 140 times lower than that of an active bat, and it burns up only 4 milligrams of fat per day. However, even small rises in the temperature of the roost dramatically increase the bats' metabolic rate. They also need to wake up periodically to drink and may move to another hibernation place.

Hibernation ends when nighttime temperatures regularly get above 50°F (10°C), when insects start to fly again in large numbers. In the southern parts of their range the bats can be active as early as February, but in northern Ontario they may not be until April.

Sperm Storage

Bats mate in the fall. Males hang in caves and make echolocation calls, and the females fly around choosing a mate. Both males and females mate with more than one partner. The females store sperm over winter in their reproductive tract, and eggs are only released and fertilized in spring. During birth the female turns around to hang from her thumbs (wing claws). She curls her body around so that as the baby is born, it drops into her tail membrane.

Mothers leave their babies clustered in nursery colonies when they fly off to feed. The young fly at 18 days, when their wings are almost adult size.

① A little brown bat hunts for insect prey. Once detected—by echolocation—the insects are snatched with the teeth in flight or scooped up with the wing tip or tail membrane.

Female little brown bats gather in nursery colonies of 50 to 5,000 individuals. The roost sites are usually in the attics of buildings. The females use the same sites year after year. If roosts are sealed, they find it difficult to locate new ones, and many probably die. While the females are gathered in their summer roosts, the males disappear, probably hiding in tree holes and other small crevices during the day.

Toward fall the males rejoin the females, and together they group in swarms to inspect hibernation sites. This happens in August in south Ontario and later in the year farther south. Hibernation sites are almost always underground in caves, mines, or even old wells. The sites need to be cold, but above freezing, ideally at 35.6°F (2°C), and with high humidity. Because their needs are so specific, good hibernation sites are rare, and bats may have to

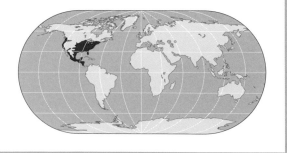

Common name Virginia opossum (common opossum, possum)

Scientific name *Didelphis virginiana*

Family	Didelphidae
Order	Didelphimorphia
Size	Length head/body: 14–22 in (35–55 cm); tail length: 10–21 in (25–54 cm)

Weight 4.5–12 lb (2–5.5 kg)

Key features Cat-sized animal with short legs and long, naked tail; pointed snout, large black eyes, and round, naked ears; white-tipped guard hairs make coat appear shaggy; hind feet have opposable first toe; female has up to 15 teats in well-developed pouch

Habits Mostly active at night; solitary; swims and climbs well

Breeding Up to 56 (usually about 21) young born after gestation period of 13 days. Young emerge from pouch at 70 days. Weaned at 3–4 months; sexually mature at 6–8 months. May live up to 5 years in captivity, 3 in the wild

Voice Clicking sounds; also growls, hisses, and screeches when angry

Diet Small animals, including reptiles, mammals, and birds; invertebrates such as insects; plant material, including fruit and leaves; carrion; human refuse

Habitat Wooded areas or scrub, usually near watercourses or close to swamps

Distribution Eastern and central U.S.; West Coast south through Central America to Nicaragua

Status Population: abundant. Widespread; increasing in numbers and range

Virginia Opossum

Didelphis virginiana

As North America's only native marsupial, the Virginia opossum is worthy of attention. However, there is far more to the species than mere geographical novelty.

THE VIRGINIA OPOSSUM IS ONE of the world's great mammalian success stories. Its ability to thrive—even where humans have made significant changes in its natural habitat—has earned it popular recognition and respect. Unlike many other marsupials, it is even known to thrive in the face of direct competition from placental mammals. Indeed, the Virginia opossum's adaptability clearly demonstrates that marsupials are not inferior to placental mammals as is sometimes implied. Opossums may lack the kind of intelligence generally attributed to monkeys or dogs, but the species compares favorably with other small- to medium-sized generalist mammals.

Disheveled Appearance

The Virginia opossum is larger than most other members of its family. It has a rather untidy, disheveled appearance owing to the long, white-tipped guard hairs that emerge beyond its soft underfur. The underfur varies from dense and woolly in the north to thin and sparse in the warmer south.

The general scruffiness of the Virginia opossum is not helped by the fact that the long tail is naked for much of its length and often damaged by frostbite, as are the ears. However, far from being just an unsightly appendage, the tail is invaluable in climbing, being fully prehensile (able to grip). Serving as a fifth limb, it can grasp branches so effectively that the opossum is able to dangle by its tail alone. The tail is also used for carrying bedding materials to the nest: The opossum scrapes leaves and grass together with its paws and then grasps the vegetation with its tail for transportation.

The Virginia opossum's feet are also highly dexterous and can grip and manipulate objects while feeding. All four feet have five toes with strong, sharp claws, except for the first toe of each hind foot, which is clawless and opposable (used like a thumb for gripping).

Healthy Appetite

Nowhere is the opossum's flexibility and generalist way of life better illustrated than in its diet. It will eat almost anything from mice, birds, and rabbits to beetles, snails, and seeds. The occasional frog or other amphibian is even known to appear on the menu. The Virginia opossum has learned the art of raiding trashcans in gardens and at picnic sites, and will happily devour food left out for birds or a pet dog. In the north of its range it takes advantage of the abundant food supplies in the summer months, often becoming very fat.

The Virginia opossum does not store food for winter, so its fat reserves can be vital in helping see it through the cold season. Even so, the animals still need to forage all year round, and they do not hibernate. Foraging generally happens at night. However, in winter the opossum may emerge more often during the day, and sometimes basks in the sun as an energy-efficient way of keeping warm.

Opossums may travel distances of up to 1.8 miles (3 km) a night in search of food, moving along the ground and through trees. They are often at least partially nomadic, occupying home ranges that cover 12 to 600 acres (5 to 240 ha). The same home range is used continuously for a period of months or years, but may be changed at any time. There can be several dens within the home range. Each consists of a nest of grass stuffed into a sheltered cavity. Typical locations include a hollow tree, a rocky crevice, a burrow, or even the corner of an attic or outhouse.

⊖ *Between leaving the pouch and achieving full independence, riding on the mother's back is the usual mode of transport for young Virginia opossums.*

⊕ *A Virginia opossum feigns death to deter a potential aggressor. The condition, which brings on symptoms akin to fainting, can last for less than a minute or as long as six hours.*

Hot-Tempered

Opossums are not territorial, but they are aggressive toward other individuals they meet. Males are especially hot-tempered; nonbreeding females less so. Rival opossums use threatening hisses and growls to warn each other off, building to harsh screeches through their 50 bared teeth. Fights are fairly common, but usually avoided by leaving regular scent marks around the home range. Such deposits provide clues to the movements of particular individuals, as well as information about their age, sex, status, and breeding condition.

Female opossums rarely live more than two years; and even if they survive into a third, they will not breed again. All their reproductive efforts must therefore be crammed into just two seasons. The breeding season is from January to August, giving most females the chance to rear two litters annually. Courtship is minimal, with every male approaching as many females as he can. In every month the female is only capable of conceiving on one or two days. If she is receptive, the pair will mate almost immediately. However, a male that approaches a nonestrous female will quickly be rejected.

Young opossums are born after just two weeks' gestation. Those that are lucky enough to reach the pouch spend a further 50 days firmly attached to a teat. Females usually have 13 teats (sometimes as many as 15). While most females give birth to about 20 babies, litters of over 50 are not unknown. Even in average-sized litters there are too many young for each to find a teat. The race to the pouch helps weed out the weakest individuals. It also ensures that the mother does not waste energy on young that are unlikely to survive. If only one or two babies reach the pouch, their feeble sucking is not enough to stimulate the production of milk. In such cases the litter is abandoned, and the whole process starts again.

Similarly, if a baby becomes detached from its teat before it can survive alone, the mother will ignore its cries, and it will die. This happens rarely, despite the inevitable crush in the pouch as the young grow and start to jostle one another. Once a young opossum starts to suck, the teat swells up inside its mouth so that it cannot let go. When the young are old enough to detach voluntarily from the teat, they accompany their mother out of the nest, riding in the pouch or on her back. They stay with her for a further month or two, becoming fully independent in just four months. Litters born in spring tend to disperse earlier than those arriving in late summer, which may stay with their mother throughout the following winter.

"Playing Possum"

Despite its reputation for aggression and belligerence, the Virginia opossum appears to know when it is beaten. Rather than try to fight an obviously superior rival or defend itself against a larger attacker, it resorts to the tactic of pretending to be dead. Various other animals employ a similar strategy, known as "playing possum." An opossum feigning death is amazingly convincing. The animal suddenly keels over on its side with its toes clenched, its body slightly curled, and its tail turned in as though it has died in the throes of some great stress or discomfort. At the same time, its mouth hangs open, and its eyes glaze over. The opossum does not flinch when handled and cannot be roused by loud noises or tickled, prodded, or shaken back to life. That deathlike state, called catatonia, can last for anything from one minute to six hours. Physiologically speaking, it is similar to fainting, but the opossum appears able to do it quite voluntarily. At the same time as "dropping dead," it allows a truly foul-smelling secretion to leak from its anus. The stench is usually enough to deter any other animal from inspecting the "corpse" too closely. The opossum waits for an opportune moment to "wake up" and make its escape.

Virginia opossums do not hibernate and so need to forage all year round. Although they prefer to shelter in bad weather, they will sometimes venture out in snow and strong winds.

High Deathrate

Female opossums born in February may have litters of their own by August, but they are the lucky ones. The deathrate among opossums is high at almost every stage of their life. In the first few days failure to attach to a teat means starvation, and as many as 10 to 25 percent of young die in the pouch. Newly independent youngsters dispersing away from their mother are easy pickings for predators, and many starve or die of exposure during their first winter. Parasite infestations are common, as are other

A Teaspoon of Babies

Newborn opossums are unbelievably tiny. Each one is smaller than a single pea and weighs about 0.007 ounces (one-fifth of a gram). Two hundred babies would weigh less than an ounce (28 g), and a litter of 20 can be carried in a teaspoon! Born after just 13 days inside their mother, they appear virtually embryonic. It seems unbelievable that such tiny, underdeveloped creatures can survive at all. Yet those 13 days have provided each baby with a heart, muscles and claws to enable it to crawl, and, somewhere in its minute brain, the will to face the immediate life-and-death challenge of reaching its mother's pouch and attaching to a teat. Doing so quickly is vital since there are rarely enough teats to go around, and babies that do not claim one will soon die.

diseases, although the species has a high level of resistance to rabies.

Roadkills account for thousands of opossum deaths every year; and as if that were not enough, many are trapped or poisoned as vermin or hunted for their fur. The fur industry peaked in the 1970s, when 1 million opossums were killed annually. Trade in opossum fur has since declined as a result of the relatively small profits to be made. At only $2 a pelt, hunting and skinning opossums is more effort than it is worth. Despite the many and varied perils facing the Virginia opossum, the animal's adaptable lifestyle and prodigious birthrate mean that the species continues to thrive.

⊖ *Infant Virginia opossums with their mother. In just 90 days they have gone from being smaller than a pea to the size shown here. But they are not safe yet. Newly independent youngsters are easy pickings for predators.*

Conquering the United States

Before European settlers arrived in North and Central America, opossums lived only in the east—ranging from Nicaragua in the south to Pennsylvania in the north. While in Australia the arrival of colonists proved disastrous for its native marsupials, in the United States the Virginia opossum took readily to life alongside humans. By the late 1950s it had spread north to New York and the shores of the Great Lakes and west into the Great Plains states of Kansas, Nebraska, and Iowa. Much of the expansion was undertaken by the opossums themselves, as they exploited the agreeable living conditions around farms and gardens. The trend was sometimes assisted by deliberate introduction of the opossum to places such as California, New England, and Ontario (Canada). Today there are opossums all the way up the West Coast of the United States and Mexico (excluding Baja California). The range of the Virginia opossum is almost double what it was only 200 years ago, and it is still growing. Not bad for a "primitive" mammal!

Glossary

Words in SMALL CAPITALS refer to other entries in the glossary.

A

Adaptation features of an animal that adjust it to its environment; may be produced by evolution—e.g., camouflage coloration

Adaptive radiation when a group of closely related animals (e.g., members of a FAMILY) have evolved differences from each other so that they can survive in different NICHES

Adult a fully grown animal that has reached breeding age

Amphibian any cold-blooded VERTEBRATE of the class Amphibia, typically living on land but breeding in the water, e.g., frogs, toads, and newts

Anal gland (anal sac) a gland opening by a short duct either just inside the anus or on either side of it

Aquatic living in water

Arboreal living among the branches of trees

Arthropod animals with a jointed outer skeleton, e.g., crabs and insects

B

Biodiversity a variety of SPECIES and the variation within them

Biomass the total weight of living material

Biped any animal that walks on two legs. See QUADRUPED

Breeding season the entire cycle of reproductive activity from courtship, pair formation (and often establishment of TERRITORY), through nesting to independence of young

Browsing feeding on leaves of trees and shrubs

C

Cache a hidden supply of food; also (verb) to hide food for future use

Callosities hardened, thickened areas on the skin (e.g., ischial callosities in some PRIMATES)

Canine (tooth) a sharp stabbing tooth usually longer than rest

Canopy continuous (closed) or broken (open) layer in forests produced by the intermingling of branches of trees

Capillaries tiny blood vessels that convey blood through organs from arteries to veins

Carnassial (teeth) opposing pair of teeth especially adapted to shear with a cutting (scissorlike) edge; in living mammals the arrangement is unique to Carnivora, and the teeth involved are the fourth upper PREMOLAR and first lower MOLAR

Carnivore meat-eating animal

Carrion dead animal matter used as a food source by scavengers

Cecum a blind sac in the digestive tract opening out from the junction between the small and large intestines. In herbivorous mammals it is often very large; it is the site of bacterial action on CELLULOSE. The end of the cecum is the appendix; in SPECIES with a reduced cecum the appendix may retain an antibacterial function

Cellulose the material that forms the cell walls of plants

Cementum hard material that coats the roots of mammalian teeth. In some SPECIES cementum is laid down in annual layers that, under a microscope, can be counted to estimate the age of individuals

Cheek pouch a pocket in or alongside the mouth used for the temporary storage of food

Cheek teeth teeth lying behind the CANINES, consisting of PREMOLARS and MOLARS

CITES Convention on International Trade in Endangered Species. An agreement between nations that restricts international trade to permitted levels through a system of licensing and administrative controls.

Cloven hoof foot that is formed from two toes, each within a horny covering

Congenital condition animal is born with

Coniferous forest evergreen forests of northern regions and mountainous areas dominated by pines, spruces, and cedars

Corm underground food storage bulb of certain plants

Crepuscular active in twilight

Cursorial adapted for running

D

Deciduous forest dominated by trees that lose their leaves in winter (or the dry season)

Deforestation the process of cutting down and removing trees for timber or to create open space for activities such as growing crops or grazing animals

Delayed implantation when the development of a fertilized egg is suspended for a variable period before it implants into the wall of the UTERUS and completes normal pregnancy. Births are thus delayed until a favorable time of year

Den a shelter, natural or constructed, used for sleeping, giving birth, and raising young; also (verb) the act of retiring to a den to give birth and raise young, or for winter shelter

Dental formula convention for summarizing the dental arrangement, in which the numbers of all types of tooth in each half of the upper and lower jaw are given. The numbers are always presented in the order: INCISOR (I), CANINE (C), PREMOLAR (P), MOLAR (M). The final figure is the total number of teeth to be found in the skull. A typical example for Carnivora is I 3/3, C1/1, P4/4, M3/3 = 44

Dentition animal's set of teeth

Desert area of low rainfall dominated by specially adapted plants such as cacti

Digit a finger or toe

Digitigrade method of walking on the toes without the heel touching the ground. See PLANTIGRADE

Dispersal the scattering of young animals going to live away from where they were born and brought up

Display any relatively conspicuous pattern of behavior that conveys specific information to others, usually to members of the same SPECIES; can involve visual or vocal elements, as in threat, courtship, or greeting displays

Diurnal active during the day

DNA (deoxyribonucleic acid) the substance that makes up the main part of the chromosomes of all living things; contains the genetic code that is handed down from generation to generation

DNA analysis "genetic fingerprinting," a technique that allows scientists to see who is related to whom, for example, which male was the father of particular offspring

Domestication process of taming and breeding animals to provide help and useful products for humans

Dorsal relating to the back or spinal part of the body; usually the upper surface

Droppings see FECES and SCATS

E

Ecosystem a whole system in which plants, animals, and their environment interact

Edentate toothless, but is also used as group name for anteaters, sloths, and armadillos

Endemic found only in one small geographical area, nowhere else

Estivation inactivity or greatly decreased activity during hot or dry weather

Estrus the period when eggs are released from the female's ovaries, and she becomes available for successful mating. Estrous females are often referred to as "in heat" or as "RECEPTIVE" to males

Eutherian mammals that give birth to babies, not eggs, and rear them without using a pouch on the mother's belly

Extinction the process of dying out in which every last individual dies, and the SPECIES is lost forever

Eyeshine when eyes of animals (especially CARNIVORES) reflect a beam of light shone at them. It is caused by a special reflective layer (the tapetum) at the back of the eye characteristic of many NOCTURNAL species and associated with an increased ability to see in the dark

F

Family technical term for group of closely related SPECIES that often also look quite similar. Zoological family names always end in "idae." Also, a social group within a species consisting of parents and their offspring

Feces remains of digested food expelled from body as pellets, often with SCENT secretions

Feral domestic animals that have gone wild and live independently of people

Flystrike where CARRION-feeding flies have laid their eggs

Fossorial adapted for digging and living in burrows or underground tunnels

Frugivore an animal that eats fruit as main part of the diet

Fur mass of hairs forming a continuous coat characteristic of mammals

Fused joined together

G

Gape wide-open mouth

Gene the basic unit of heredity enabling one generation to pass on characteristics to its offspring

Generalist an animal that is capable of a wide range of activities, not specialized

Genus a group of closely related SPECIES. The plural is genera.

Gestation the period of pregnancy between fertilization of the egg and birth of the baby

Grazing feeding on grass

Gregarious living together in loose groups or herds

H

Harem a group of females living in the same TERRITORY and consorting with a single male

Herbivore an animal that eats plants (grazers and browsers are thus herbivores)

Heterodont DENTITION specialized into CANINES, INCISORS, and PREMOLARS, each type of tooth having a different function. See HOMODONT

Hibernation becoming inactive in winter, with lowered body temperature to save energy. Hibernation takes place in a special nest or DEN called a hibernaculum

Homeothermy maintenance of a high and constant body temperature by means of internal processes; also called "warm-blooded"

Home range the area that an animal uses in the course of its normal periods of activity. See TERRITORY

Homodont DENTITION in which the teeth are all similar in appearance and function

Horns a pair of sharp, unbranched prongs projecting from the head of CLOVEN-HOOFED animals. Horns have a bony core with a tough outer covering made of KERATIN like fingernails

Hybrid offspring of two closely related SPECIES that can interbreed, but the hybrid is sterile

I

Inbreeding breeding among closely related animals (e.g., cousins) leading to weakened genetic composition and reduced survival rates

Incisor (teeth) simple pointed teeth at the front of the jaws used for nipping and snipping

Indigenous living naturally in a region; NATIVE (i.e., not an introduced SPECIES)

Insectivore animals that feed on insects and similar small prey. Also used as a group name for animals such as hedgehogs, shrews, and moles

Interbreeding breeding between animals of different SPECIES or varieties within a single FAMILY or strain; interbreeding can cause dilution of the gene pool

Interspecific between SPECIES

Intraspecific between individuals of the same SPECIES

Invertebrates animals that have no backbone (or other true bones) inside their body, e.g., mollusks, insects, jellyfish, and crabs

IUCN CATEGORIES

EX Extinct, when there is no reasonable doubt that the last individual of a species has died.

EW Extinct in the Wild, when a species is known only to survive in captivity or as a naturalized population well outside the past range.

CR Critically Endangered, when a species is facing an extremely high risk of extinction in the wild in the immediate future.

EN Endangered, when a species faces a very high risk of extinction in the wild in the near future.

VU Vulnerable, when a species faces a high risk of extinction in the wild in the medium-term future.

LR Lower Risk, when a species has been evaluated and does not satisfy the criteria for CR, EN, or VU.

DD Data Deficient, when there is not enough information about a species to assess the risk of extinction.

NE Not Evaluated, species that have not been assessed by the IUCN criteria.

IUCN International Union for the Conservation of Nature, responsible for assigning animals and plants to internationally agreed categories of rarity. See table above

J

Juvenile young animal that has not yet reached breeding age

K

Kelp brown seaweeds

Keratin tough, fibrous material that forms hairs, feathers, and protective plates on the skin of VERTEBRATE animals

L

Lactation process of producing milk in MAMMARY GLANDS for offspring

Larynx voice box where sounds are created

Latrine place where FECES are left regularly, often with SCENT added

Leptospirosis disease caused by leptospiral bacteria in kidneys and transmitted via urine

M

Mammary glands characteristic of mammals, glands for production of milk

Marine living in the sea

Matriarch senior female member of a social group

Metabolic rate the rate at which chemical activities occur within animals, including the exchange of gasses in respiration and the liberation of energy from food

Metabolism the chemical activities within animals that turn food into energy

Migration movement from one place to another and back again, usually seasonal

Molars large crushing teeth at the back of the mouth

Molt process in which mammals shed hair, usually seasonal

Monogamous animals that have only one mate at a time

Montane in a mountain environment

Musk mammalian SCENT

Mutation random changes in genetic material

N

Native belonging to that area or country, not introduced by human assistance

Natural selection when animals and plants are challenged by natural processes (including predation and bad weather) to ensure survival of the fittest

New World the Americas; OLD WORLD refers to the non-American continents (not usually Australia)

Niche part of a habitat occupied by an ORGANISM, defined in terms of all aspects of its lifestyle

Nocturnal active at night

Nomadic animals that have no fixed home, but wander continuously

O

Old World non-American continents. See NEW WORLD

Olfaction sense of smell

Omnivore an animal that eats almost anything, meat or vegetable

Opportunistic taking advantage of every varied opportunity that arises, flexible behavior

Opposable fingers or toes that can be brought to bear against others on the same hand or foot in order to grip objects

Order a subdivision of a class of animals consisting of a series of related animal FAMILIES.

Ovulation release of egg from the female's ovary prior to its fertilization

P

Pair bond behavior that keeps a male and a female together beyond the time it takes to mate; marriage is a "pair bond"

Parasite animal or plant that lives on or in body of another

Parturition process of giving birth

Pelage the furry coat of a mammal

Pelt furry coat; often refers to skin removed from animal as fur

Pheromone SCENT produced by animals to enable others to find and recognize them

Physiology the processes and workings within plants and animal bodies, e.g., digestion. Keeping a warm-blooded state is a part of mammal physiology

Placenta the structure that links an embryo to its mother during pregnancy, allowing exchange of chemicals between them

Plantigrade walking on the soles of the feet with the heels touching the ground. See DIGITIGRADE

Polygamous when animals have more than one mate in a single mating season. MONOGOMOUS animals have only a single mate

Polygynous when a male mates with several females in one BREEDING SEASON

Population a distinct group of animals of the same SPECIES or all the animals of that species

Posterior the hind end or behind another structure

Predator an animal that kills live prey for food

Prehensile grasping tail or fingers

Premolars teeth found in front of MOLARS, but behind CANINES

Pride social group of lions

Primate a group of mammals that includes monkeys, apes, and ourselves

Promiscuous mating often with many mates, not just one

Protein chemicals made up of amino acids. Essential in the diet of animals

Q

Quadruped an animal that walks on all fours (a BIPED walks on two legs)

R

Range the total geographical area over which a SPECIES is distributed

Receptive when a female is ready to mate (in ESTRUS)

Reproduction the process of breeding, creating new offspring for the next generation

Retina light-sensitive layer at the back of the eye. May include a tapetum, a reflective layer causing EYESHINE when a beam of light is shone at the eyes

Retractile capable of being withdrawn, as in the claws of typical cats, which can be folded back into the paws to protect from damage when walking

Riparian living beside rivers and lakes

Roadkill animals killed by road traffic

Rumen complex stomach found in RUMINANTS specifically for digesting plant material

Ruminant animals that eat vegetation and later bring it back from the stomach to chew again ("chewing the cud" or "rumination") to assist its digestion by microbes in the stomach

S

Savanna tropical grasslands with scattered trees and low rainfall, usually in warm areas

Scats fecal pellets, especially of CARNIVORES. SCENT is often deposited with the pellets as territorial markers

Scent chemicals produced by animals to leave smell messages for others to find and interpret

Scrotum bag of skin within which the male testicles are located

Scrub vegetation that is dominated by shrubs—woody plants usually with more than one stem

Secondary forest trees that have been planted or grown up on cleared ground

Siblings brothers and sisters

Social behavior interactions between individuals within the same SPECIES, e.g., courtship

Species a group of animals that look similar and can breed to produce fertile offspring

Spraint hunting term for SCATS (see above) of certain CARNIVORES, especially otters

Steppe open grassland in parts of the world where the climate is too harsh for trees to grow

Sub-Saharan all parts of Africa lying south of the Sahara Desert

Subspecies a locally distinct group of animals that differ slightly from the normal appearance of the SPECIES; often called a race

Symbiosis when two or more SPECIES live together for their mutual benefit more successfully than either could live on its own

T

Taxonomy the branch of biology concerned with classifying ORGANISMS into groups according to similarities in their structure, origins, or behavior. The categories, in order of increasing broadness, are: SPECIES, GENUS, FAMILY, ORDER, class, and phylum.

Terrestrial living on land

Territory defended space

Thermoregulation the maintenance of a relatively constant body temperature either by adjustments to METABOLISM or by moving between sunshine and shade

Torpor deep sleep accompanied by lowered body temperature and reduced METABOLIC RATE

Translocation transferring members of a SPECIES from one location to another

Tundra open grassy or shrub-covered lands of the far north

U

Underfur fine hairs forming a dense, woolly mass close to the skin and underneath the outer coat of stiff hairs in mammals

Ungulate hoofed animals such as pigs, deer, cattle, and horses; mostly HERBIVORES

Uterus womb in which embryos of mammals develop

V

Ventral the belly or underneath of an animal (opposite of DORSAL)

Vertebrate animal with a backbone (e.g., fish, mammals, reptiles), usually with a skeleton made of bones, but sometimes softer cartilage

Vibrissae sensory whiskers, usually on snout, but can be on areas such as elbows, tail, or eyebrows

Viviparous animals that give birth to active young rather than laying eggs

Vocalization making of sounds such as barking and croaking

Z

Zoologist person who studies animals

Zoology the study of animals

Further Reading

Buskirk, S. W. **Martens, Sables, and Fishers: Biology and Conservation**. Ithaca, NY: Cornell University Press, 1994.

Byers, J. A. **American Pronghorn**. Chicago: University of Chicago Press, 1997.

Carwardine, M. **Whales, Dolphins, and Porpoises**. New York: Dorling Kindersley, 1995.

Corrigan, P. **The Whale-Watcher's Guide**. Minnetonka, WI: NorthWord Press, 1994.

Craves, R. A. **The Prairie Dog: Sentinel of the Plains**. Austin, TX: Texas Technical University Press, 2001.

De La Rosa, C., and C. Nocke. **A Guide to the Carnivores of Central America**. Austin, TX: University of Texas Press, 2000.

Fenton, M. B. **Bats**. New York: Facts on File, 1992.

Furtman, M. **Seasons of the Elk**. Minnetonka, WI: NorthWord Press, 1997.

Gaskin, D. E. **The Ecology of Whales and Dolphins**. Portsmouth, NH: Heinemann, 1982.

Hart, M. **Rats**. New York: Allison and Busby/Schocken Books, 1982.

Isenberg, A. **The Destruction of the Bison**. Cambridge, UK: Cambridge University Press, 2000.

Jones, K. **Wolf Mountains**. Calgary: Calgary University Press, 2002.

Kaufman, G. D., and P. H. Forestell. **Hawaii's Humpback Whales**. Maui, HI: Pacific Whale Foundation Press, 1986.

Kitchener, A. **The Natural History of Wildcats**. Ithaca, NY: Cornell University Press, 1991.

Lacy, E. A., J. L. Patton, and G. N. Cameron. **Life Underground: The Biology of Subterranean Rodents**. Chicago: Chicago University Press, 2000.

Leatherwood, S., and R. R. Reeves. **The Sierra Club Handbook of Whales and Dolphins of the World**. San Francisco: Sierra Club, 1983.

MacDonald, D. **The Encyclopedia of Mammals**. New York: Barnes and Noble, 2001.

McIntyre, R. **War Against the Wolf**. Stillwater, MN: Voyageur Press, 1995.

Mech, L. D. **The Wolf: Ecology and Behavior of an Endangered Species**. Minneapolis: Minnesota University Press, 1981.

Neuweiler, G. **The Biology of Bats**. New York: Oxford University Press, 2000.

Nowak, R. M. **Walker's Bats of the World**. Baltimore: John Hopkins University Press, 1995.

———. **Walker's Mammals of the World**. Baltimore: John Hopkins University Press, 1999.

Payne, R. **Among Whales**. New York: Bantam Press, 1996.

Perrin, W. **Encyclopedia of Marine Mammals**. New York: Academic Press, 2002.

Reynolds, J. E. **The Bottlenose Dolphin: Biology and Conservation**. Tallahassee, FL: Florida University Press, 2000.

Ripple, J., and D. Perrine. **Manatees and Dugongs of the World**. Stillwater, MN: Voyageur Press, 2001.

Servheen, C. **Bears: Status Survey and Conservation Action Plan**. Cambridge, UK: IUCN, 1999.

Steele, M., and J. L. Koprowski. **North American Tree squirrels**. Washington, DC: Smithsonian Institution Press, 2001.

Sunquist, M., and F. Sunquist. **Wildcats of the World**. Chicago: Chicago University Press, 2002.

Twiss, J. R., and R. R. Reeves. **Conservation and Management of Marine Mammals**. Washington, DC: Smithsonian Institution Press, 1999.

Wells-Gosling, N. **Flying Squirrels**. Washington, DC: Smithsonian Institution Press, 1985.

Whitaker, J. O. **National Audubon Society Field Guide to North American Mammals**. New York: Alfred A. Knopf, 1996.

Wilson, D. E. **The Smithsonian Book of North American Mammals**. Washington, DC: Smithsonian Institution Press, 1999.

Wilson, D. E., and D. M. Reeder. **Mammal Species of the World. A Taxonomic and Geographical Reference**. Washington, DC: Smithsonian Institution Press, 1999.

Young, J. Z. **The Life of Mammals: Their Anatomy and Physiology**. Oxford, UK: Oxford University Press, 1975.

Useful Websites

General

http://animaldiversity.ummz.umich.edu/
University of Michigan Museum of Zoology animal diversity websites. Search for pictures and information about animals by class, family, and common name. Includes glossary

http://www.cites.org/
IUCN and CITES listings. Search for animals by scientific name, order, family, genus, species, or common name. Location by country and explanation of reasons for listings

http://endangered.fws.gov
Information about threatened animals and plants from the U.S. Fish and Wildlife Service, the organization in charge of 94 million acres (38 million ha) of American wildlife refuges

http://www.iucn.org
Details of species and their status; listings by the International Union for the Conservation of Nature, also lists IUCN publications

http://www.panda.org
World Wide Fund for Nature (WWF), newsroom, press releases, government reports, campaigns

http://www.aza.org
American Zoo and Aquarium Association

http://www.wcs.org
Website of the Wildlife Conservation Society

http://www.nwf.org
Website of the National Wildlife Federation

http://www.nmnh.si.edu/msw/
Mammals list on Smithsonian Museum site

http://www.press.jhu.edu/books/walkers_mammals_of_the_world/prep.html
Text of basic book listing species, illustrating almost every genus

Specific Groups

http://www.defenders.org/
Active conservation of carnivores, including wolves and grizzly bears

http://www.wwfcanada.org/en/res_links/pdf/projdesc.pdf
Carnivore conservation in the Rocky Mountains

http://www.savethemanatee.org/
Save the manatee club; adopt one if you want

http://home.t-online.de/home/rothauscher/dugong.htm
News and links about dugongs

http://www.sirenian.org
General information about sirenians

http://www.whalecenter.org/
The Whale Center of New England

http://www.cetaceanresearch.com/sounds.html
Underwater sounds of killer whales

http://www.acsonline.org/
American Cetacean Society; supports and reports research and conservation

http://www.pinnipeds.org
General website for seals and sea lions

http://www.ultimateungulate.com
Guide to world's hoofed mammals

http://www.wildhorseandburro.blm.gov/Wild horses and burros in North America

http://www.batcon.org
Bat Conservation International

http://www.desertusa.com/jan97/du_bats.html
Desert bats (U.S.)

http://www.tpwd.state.tx.us/nature/wild/mammals/dillo.htm
Nine-banded armadillo

spot.colorado.edu/~halloran/sqrl.html
Tree squirrels in North America

http://www.rmca.org/
Rat and Mouse Club of America. Includes club information, events, contests, merchandise, and rat standards

http://www.alienexplorer.com/ecology/topic24.html
Squirrels and chipmunks

http://www.ngpc.state.ne.us/wildlife/flysqu.html
Habits, reproduction, and management of flying squirrels

http://www.deadsquirrel.com
Squirrels as pests, fun too!

http://www.enchantedlearning.com/subjects/mammals/lagomorphs/
General website covering Lagomorphs

http://www.desertusa.com/july96/du_rabbi.html
Jackrabbit information

http://www.geobop.com/Mammals/Lagomorpha/
General information about rabbits, hares, and pikas

http://www.desertusa.com/mag99/mar/papr/porcupine.html
Information on North American porcupines

Index

Picture Credits

Artists